The publishing house tredition has created the series **TREDITION CLASSICS**. It contains classical literature works from over two thousand years. Most of these titles have been out of print and off the bookstore shelves for decades.

The book series is intended to preserve the cultural legacy and to promote the timeless works of classical literature. As a reader of a **TREDITION CLASSICS** book, the reader supports the mission to save many of the amazing works of world literature from oblivion.

The symbol of **TREDITION CLASSICS** is Johannes Gutenberg (1400 – 1468), the inventor of movable type printing.

With the series, tredition intends to make thousands of international literature classics available in printed format again – worldwide.

All books are available at book retailers worldwide in paperback and in hardcover. For more information please visit: www.tredition.com

tredition was established in 2006 by Sandra Latusseck and Soenke Schulz. Based in Hamburg, Germany, tredition offers publishing solutions to authors and publishing houses, combined with worldwide distribution of printed and digital book content. tredition is uniquely positioned to enable authors and publishing houses to create books on their own terms and without conventional manufacturing risks.

For more information please visit: www.tredition.com

The Story of Germ Life

H. W. (Herbert William) Conn

Imprint

This book is part of the TREDITION CLASSICS series.

Author: H. W. (Herbert William) Conn
Cover design: toepferschumann, Berlin (Germany)

Publisher: tredition GmbH, Hamburg (Germany)
ISBN: 978-3-8495-0807-4

www.tredition.com
www.tredition.de

Copyright:
The content of this book is sourced from the public domain.

The intention of the TREDITION CLASSICS series is to make world literature in the public domain available in printed format. Literary enthusiasts and organizations worldwide have scanned and digitally edited the original texts. tredition has subsequently formatted and redesigned the content into a modern reading layout. Therefore, we cannot guarantee the exact reproduction of the original format of a particular historic edition. Please also note that no modifications have been made to the spelling, therefore it may differ from the orthography used today.

PREFACE.

Since the first edition of this book was published the popular idea of bacteria to which attention was drawn in the original preface has undergone considerable modification. Experimental medicine has added constantly to the list of diseases caused by bacterial organisms, and the general public has been educated to an adequate conception of the importance of the germ as the chief agency in the transmission of disease, with corresponding advantage to the efficiency of personal and public hygiene. At the same time knowledge of the benign bacteria and the enormous role they play in the industries and the arts has become much more widely diffused. Bacteriology is being studied in colleges as one of the cultural sciences; it is being widely adopted as a subject of instruction in high schools; and schools of agriculture and household science turn out each year thousands of graduates familiar with the functions of bacteria in daily life. Through these agencies the popular misconception of the nature of micro- organisms and their relations to man is being gradually displaced by a general appreciation of their manifold services. It is not unreasonable to hope that the many thousands of copies of this little manual which have been circulated and read have contributed materially to that end. If its popularity is a safe criterion, the book has amply fulfilled its purpose of placing before the general reader in a simple and direct style the main facts of bacteriology. Beginning with a discussion of the nature of bacteria, it shows their position in the scale of plant and animal life. The middle chapters describe the functions of bacteria in the arts, in the dairy, and in agriculture. The final chapters discuss the relation of bacteria to disease and the methods by which the new and growing science of preventive medicine combats and counteracts their dangerous powers.

JULY, 1915.

CONTENTS.

I.—BACTERIA AS PLANTS

Historical.—Form of bacteria.—Multiplication of bacteria.—Spore formation.—Motion.—Internal structure.—Animals or plants?—Classification.—Variation.—Where bacteria are found.

II.—MISCELLANEOUS USES OF BACTERIA IN THE ARTS.

Maceration industries.—Linen.—Jute.—Hemp.—Sponges.—Leather.
—Fermentative industries.—Vinegar—Lactic acid.—Butyric acid.—Bacteria in tobacco curing.—Troublesome fermentations.

III.—BACTERIA IN THE DAIRY.

Sources of bacteria in milk.—Effect of bacteria on milk.—Bacteria used in butter making.—Bacteria in cheese making.

IV.—BACTERIA IN NATURAL PROCESSES.

Bacteria as scavengers.—Bacteria as agents in Nature's food cycle.—Relation of bacteria to agriculture.—Sprouting of seeds. —The silo.—The fertility of the soil.—Bacteria as sources of trouble to the farmer.—Coal formation.

V.—PARASITIC BACTERIA AND THEIR RELATION TO DISEASE

Method of producing disease.—Pathogenic germs not strictly parasitic.—Pathogenic germs that are true parasites.—What diseases are due to bacteria.—Variability of pathogenic powers.—Susceptibility of the individual.—Recovery from bacteriological diseases.—Diseases caused by organisms other than bacteria.

VI. – METHODS OF COMBATING PARASITIC BACTERIA

Preventive medicine. – Bacteria in surgery. – Prevention by inoculation. – Limits of preventive medicine. – Curative medicine. – Drugs – Vis medicatrix naturae. – Antitoxines and their use. – Conclusion.

THE STORY OF GERM LIFE.

CHAPTER I.

BACTERIA AS PLANTS.

During the last fifteen years the subject of bacteriology [Footnote: The term microbe is simply a word which has been coined to include all of the microscopic plants commonly included under the terms bacteria and yeasts.] has developed with a marvellous rapidity. At the beginning of the ninth decade of the century bacteria were scarcely heard of outside of scientific circles, and very little was known about them even among scientists. Today they are almost household words, and everyone who reads is beginning to recognise that they have important relations to his everyday life. The organisms called bacteria comprise simply a small class of low plants, but this small group has proved to be of such vast importance in its relation to the world in general that its study has little by little crystallized into a science by itself. It is a somewhat anomalous fact that a special branch of science, interesting such a large number of people, should be developed around a small group of low plants. The importance of bacteriology is not due to any importance bacteria have as plants or as members of the vegetable kingdom, but solely to their powers of producing profound changes in Nature. There is no one family of plants that begins to compare with them in importance. It is the object of this work to point out briefly how much both of good and ill we owe to the life and growth of these microscopic organisms. As we have learned more and more of them during the last fifty years, it has become more and more evident that this one little class of microscopic plants fills a place in Nature's processes which in some respects balances that filled by the whole of the green plants. Minute as they are, their importance can hardly be overrated, for upon their activities is founded the continued life of the animal and vegetable kingdom. For good and for ill they are agents of neverceasing and almost unlimited powers.

HISTORICAL.

The study of bacteria practically began with the use of the microscope. It was toward the close of the seventeenth century that the Dutch microscopist, Leeuwenhoek, working with his simple lenses, first saw the organisms which we now know under this name, with sufficient clearness to describe them. Beyond mentioning their existence, however, his observations told little or nothing. Nor can much more be said of the studies which followed during the next one hundred and fifty years. During this long period many a microscope was turned to the observation of these minute organisms, but the majority of observers were contented with simply seeing them, marvelling at their minuteness, and uttering many exclamations of astonishment at the wonders of Nature. A few men of more strictly scientific natures paid some attention to these little organisms. Among them we should perhaps mention Von Gleichen, Muller, Spallanzani, and Needham. Each of these, as well as others, made some contributions to our knowledge of microscopical life, and among other organisms studied those which we now call bacteria. Speculations were even made at these early dates of the possible causal connection of these organisms with diseases, and for a little the medical profession was interested in the suggestion. It was impossible then, however, to obtain any evidence for the truth of this speculation, and it was abandoned as unfounded, and even forgotten completely, until revived again about the middle of the 19th century. During this century of wonder a sufficiency of exactness was, however, introduced into the study of microscopic organisms to call for the use of names, and we find Muller using the names of Monas, Proteus, Vibrio, Bacillus, and Spirillum, names which still continue in use, although commonly with a different significance from that given them by Muller. Muller did indeed make a study sufficient to recognise the several distinct types, and attempted to classsify these bodies. They were not regarded as of much importance, but simply as the most minute organisms known.

Nothing of importance came from this work, however, partly because of the inadequacy of the microscopes of the day, and partly because of a failure to understand the real problems at issue. When we remember the minuteness of the bacteria, the impossibility of studying any one of them for more than a few moments at a time — only so long, in fact, as it can be followed under a microscope; when we remember, too, the imperfection of the compound microscopes which made high powers practical impossibilities; and, above all, when we appreciate the looseness of the ideas which pervaded all scientists as to the necessity of accurate observation in distinction from inference, it is not strange that the last century gave us no knowledge of bacteria beyond the mere fact of the existence of some extremely minute organisms in different decaying materials. Nor did the 19th century add much to this until toward its middle. It is true that the microscope was vastly improved early in the century, and since this improvement served as a decided stimulus to the study of microscopic life, among other organisms studied, bacteria received some attention. Ehrenberg, Dujardin, Fuchs, Perty, and others left the impress of their work upon bacteriology even before the middle of the century. It is true that Schwann shrewdly drew conclusions as to the relation of microscopic organisms to various processes of fermentation and decay — conclusions which, although not accepted at the time, have subsequently proved to be correct. It is true that Fuchs made a careful study of the infection of "blue milk," reaching the correct conclusion that the infection was caused by a microscopic organism which he discovered and carefully studied. It is true that Henle made a general theory as to the relation of such organisms to diseases, and pointed out the logically necessary steps in a demonstration of the causal connection between any organism and a disease. It is true also that a general theory of the production of ail kinds of fermentation by living organisms had been advanced. But all these suggestions made little impression. On the one hand, bacteria were not recognised as a class of organisms by themselves — were not, indeed, distinguished from yeasts or other minute animalcuise. Their variety was not mistrusted and their significance not conceived. As microscopic organisms, there were no reasons for considering them of any more importance than any other small animals or plants, and their extreme minuteness and simplicity made them of little interest to the microscopist. On the other hand, their causal connection with fermentative and putrefactive processes was entirely obscured by the overshadowing weight of the chemist Liebig, who believed that fermentations and putrefactions were simply chemical processes. Liebig insisted that all albuminoid bodies were in a state of chemically unstable equilibrium, and if left to themselves would fall to pieces without any need of the action of microscopic organisms. The force of Liebig's authority and the brilliancy of his expositions led to the wide acceptance of his views and the temporary obscurity of the relation of microscopic organisms to fermentative and putrefactive processes. The objections to Liebig's views were hardly noticed, and the force of the experiments of Schwann was silently ignored. Until the sixth decade of the century, therefore, these organisms, which have since become the basis of a new branch of science, had hardly emerged from obscurity. A few microscopists recognised their existence, just as they did any other group of small animals or plants, but even yet they failed to look upon them as forming a distinct group. A growing number of observations was accumulating, pointing toward a probable causal connection between fermentative and putrefactive processes and the growth of microscopic organisms; but these observations were known only to a few, and were ignored by the majority of scientists.

It was Louis Pasteur who brought bacteria to the front, and it was by his labours that these organisms were rescued from the obscurity of scientific publications and made objects of general and crowning interest. It was Pasteur who first successfully combated the chemical theory of fermentation by showing that albuminous matter had no inherent tendency to decomposition. It was Pasteur who first clearly demonstrated that these little bodies, like all larger animals and plants, come into existence only by ordinary methods of reproduction, and not by any spontaneous generation, as had been earlier claimed. It was Pasteur who first proved that such a common phenomenon as. the souring of milk was produced by microscopic organisms growing in the milk. It was Pasteur who first succeeded in demonstrating that certain species of microscopic organisms are the cause of certain diseases, and in suggesting successful methods of avoiding them. All these discoveries were made in rapid succession. Within ten years of the time that his name began to be heard in this connection by scientists, the subject had advanced so rapidly that it had become evident that here was a new subject of importance to the scientific world, if not to the public at large. The other important discoveries which Pasteur made it is not our purpose to mention here. His claim to be considered the founder of bacteriology will be recognised from what has already been mentioned. It was not that he first discovered the organisms, or first studied them; it was not that he first suggested their causal connection with fermentation and disease, but it was because he for the first time placed the subject upon a firm foundation by proving with rigid experiment some of the suggestions made by others, and in this way turned the attention of science to the study of micro-organisms.

After the importance of the subject had been demonstrated by Pasteur, others turned their attention in the same direction, either for the purpose of verification or refutation of Pasteur's views. The advance was not very rapid, however, since bacteriological experimentation proved to be a subject of extraordinary difficulty. Bacteria were not even yet recognised as a group of organisms distinct enough to be grouped by themselves, but were even by Pasteur at first confounded with yeasts. As a distinct group of organisms they were first distinguished by Hoffman in 1869, since which date the term bacteria, as applying to this special group of organisms, has been coming more and more into use. So difficult were the investigations, that for years there were hardly any investigators besides Pasteur who could successfully handle the subject and reach conclusions which could stand the test of time. For the next thirty years, although investigators and investigations continued to increase, we can find little besides dispute and confusion along this line. The difficulty of obtaining for experiment any one kind of bacteria by itself, unmixed with others (pure cultures), rendered advance almost impossible. So conflicting were the results that the whole subject soon came into almost hopeless confusion, and very few steps were taken upon any sure basis. So difficult were the methods, so contradictory and confusing the results, because of impure cultures, that a student of to-day who wishes to look up the previous discoveries in almost any line of bacteriology need hardly go back of 1880, since he can almost rest assured that anything done earlier than that was more likely to be erroneous than correct.

The last fifteen years have, however, seen a wonderful change. The difficulties had been mostly those of methods of work, and with the ninth decade of the century these methods were simplified by Robert Koch. This simplification of method for the first time placed this line of investigation within the reach of scientists who did not have the genius of Pasteur. It was now possible to get pure cultures easily, and to obtain with such pure cultures results which were uniform and simple. It was now possible to take steps which had the stamp of accuracy upon them, and which further experiment did not disprove. From the time when these methods were thus made manageable the study of bacteria increased with a rapidity which has been fairly startling, and the information which has accumulated is almost formidable. The very rapidity with which the investigations have progressed has brought considerable confusion, from the fact that the new discoveries have not had time to be properly assimilated into knowledge. Today many facts are known whose significance is still uncertain, and a clear logical discussion of the facts of modern bacteriology is not possible. But sufficient knowledge has been accumulated and digested to show us at least the direction along which bacteriological advance is tending, and it is to the pointing out of these directions that the following pages will be devoted.

WHAT ARE BACTERIA?

The most interesting facts connected with the subject of bacteriology concern the powers and influence in Nature possessed by the bacteria. The morphological side of the subject is interesting enough to the scientist, but to him alone. Still, it is impossible to attempt to study the powers of bacteria without knowing something of the organisms themselves. To understand how they come to play an important part in Nature's processes, we must know first how they look and where they are found. A short consideration of certain morphological facts will therefore be necessary at the start.

FORM OF BACTERIA.

In shape bacteria are the simplest conceivable structures. Although there are hundreds of different species, they have only three general forms, which have been aptly compared to billiard balls, lead pencils, and corkscrews. Spheres, rods, and spirals represent all shapes. The spheres may be large or small, and may group themselves in various ways; the rods may be long or short, thick or slender; the spirals may be loosely or tightly coiled, and may have only one or two or may have many coils, and they may be flexible or stiff; but still rods, spheres, and spirals comprise all types.

In size there is some variation, though not very great. All are extremely minute, and never visible to the naked eye. The spheres vary from 0.25 u to 1.5 u (0.000012 to 0.00006 inches). The rods may be no more than 0.3 u in diameter, or may be as wide as 1.5 u to 2.5 u, and in length vary all the way from a length scarcely longer than their diameter to long threads. About the same may be said of the spiral forms. They are decidedly the smallest living organisms which our microscopes have revealed.

In their method of growth we find one of the most characteristic features. They universally have the power of multiplication by simple division or fission. Each individual elongates and then divides in the middle into two similar halves, each of which then repeats the process. This method of multiplication by simple division is the distinguishing mark which separates the bacteria from the yeasts, the latter plants multiplying by a process known as budding. Fig. 2 shows these two methods of multiplication.

While all bacteria thus multiply by division, certain differences in the details produce rather striking differences in the results. Considering first the spherical forms, we find that some species divide, as described, into two, which separate at once, and each of which in turn divides in the opposite direction, called Micrococcus, (Fig. 3). Other species divide only in one direction. Frequently they do not separate after dividing, but remain attached. Each, however, again elongates and divides again, but all still remain attached. There are thus formed long chains of spheres like strings of beads, called Streptococci (Fig. 4). Other species divide first in one direction, then at right angles to the first division, and a third division follows at right angles to the plane of the first two, thus producing solid groups of fours, eights, or sixteens (Fig 5), called Sarcina. Each different species of bacteria is uniform in its method of division, and these differences are therefore indications of differences in species, or, according to our present method of classification, the different methods of division represent different genera. All bacteria producing Streptococcus chains form a single genus Streptococcus, and all which divide in three division planes form another genus, Sarcina, etc.

The rod-shaped bacteria also differ somewhat, but to a less extent. They almost always divide in a plane at right angles to their longest dimension. But here again we find some species separating immediately after division, and thus always appearing as short rods (Fig. 6), while others remain attached after division and form long chains. Sometimes they appear to continue to increase in length without showing any signs of division, and in this way long threads are formed (Fig. 7). These threads are, however, potentially at least, long chains of short rods, and under proper conditions they will break up into such short rods, as shown in Fig. 7a. Occasionally a rod species may divide lengthwise, but this is rare. Exactly the same may be said of the spiral forms. Here, too, we find short rods and long chains, or long spiral filaments in which can be seen no division into shorter elements, but which, under certain conditions, break up into short sections.

RAPIDITY OF MULTIPLICATION.

It is this power of multiplication by division that makes bacteria agents of such significance. Their minute size would make them harmless enough if it were not for an extraordinary power of multiplication. This power of growth and division is almost incredible. Some of the species which have been carefully watched under the microscope have been found under favourable conditions to grow so rapidly as to divide every half hour, or even less. The number of offspring that would result in the course of twenty-four hours at this rate is of course easily computed. In one day each bacterium would produce over 16,500,000 descendants, and in two days about 281,500,000,000. It has been further calculated that these 281,500,000,000 would form about a solid pint of bacteria and weigh about a pound. At the end of the third day the total descendants would amount to 47,000,000,000,000, and would weigh about 16,000,000 pounds. Of course these numbers have no significance, for they are never actual or even possible numbers. Long before the offspring reach even into the millions their rate of multiplication is checked either by lack of food or by the accumulation of their own excreted products, which are injurious to them. But the figures do have interest since they show faintly what an unlimited power of multiplication these organisms have, and thus show us that in dealing with bacteria we are dealing with forces of almost infinite extent.

This wonderful power of growth is chiefly due to the fact that bacteria feed upon food which is highly organized and already in condition for absorption. Most plants must manufacture their own foods out of simpler substances, like carbonic dioxide (CO_2) and water, but bacteria, as a rule, feed upon complex organic material already prepared by the previous life of plants or animals. For this reason they can grow faster than other plants. Not being obliged to make their own foods like most plants, nor to search for it like animals, but living in its midst, their rapidity of growth and multiplication is limited only by their power to seize and assimilate this food. As they grow in such masses of food, they cause certain chemical changes to take place in it, changes doubtless directly connected with their use of the material as food. Recognising that they do cause chemical changes in food material, and remembering this marvellous power of growth, we are prepared to believe them capable of producing changes wherever they get a foothold and begin to grow. Their power of feeding upon complex organic food and producing chemical changes therein, together with their marvellous power of assimilating this material as food, make them agents in Nature of extreme importance.

DIFFERENCES BETWEEN DIFFERENT SPECIES OF BACTERIA.

While bacteria are thus very simple in form, there are a few other slight variations in detail which assist in distinguishing them. The rods are sometimes very blunt at the ends, almost as if cut square across, while in other species they are more rounded and occasionally slightly tapering. Sometimes they are surrounded by a thin layer of some gelatinous substance, which forms what is called a capsule (Fig. 10). This capsule may connect them and serve as a cement, to prevent the separate elements of a chain from falling apart.

Sometimes such a gelatinous secretion will unite great masses of bacteria into clusters, which may float on the surface of the liquid in which they grow or may sink to the bottom. Such masses are called zoogloea, and their general appearance serves as one of the characters for distinguishing different species of bacteria (Fig. 10, a and b). When growing in solid media, such as a nutritious liquid made stiff with gelatine, the different species have different methods of spreading from their central point of origin. A single bacterium in the midst of such a stiffened mass will feed upon it and produce descendants rapidly; but these descendants, not being able to move through the gelatine, will remain clustered together in a mass, which the bacteriologist calls a colony. But their method of clustering, due to different methods of growth, is by no means always alike, and these colonies show great differences in general appearance. The differences appear to be constant, however, for the same species of bacteria, and hence the shape and appearance of the colony enable bacteriologists to discern different species (Fig. II). All these points of difference are of practical use to the bacteriologist in distinguishing species.

SPORE FORMATION.

In addition to their power of reproduction by simple division, many species of bacteria have a second method by means of spores. Spores are special rounded or oval bits of bacteria protoplasm capable of resisting adverse conditions which would destroy the ordinary bacteria. They arise among bacteria in two different methods.

Endogenous spores.—These spores arise inside of the rods or the spiral forms (Fig. 12). They first appear as slight granular masses, or as dark points which become gradually distinct from the rest of the rod. Eventually there is thus formed inside the rod a clear, highly refractive, spherical or oval spore, which may even be of a greater diameter than the rod producing it, thus causing it to swell out and become spindle formed [Fig. 12 c]. These spores may form in the middle or at the ends of the rods (Fig. 12). They may use up all the protoplasm of the rod in their formation, or they may use only a small part of it, the rod which forms them continuing its activities in spite of the formation of the spores within it. They are always clear and highly refractive from containing little water, and they do not so readily absorb staining material as the ordinary rods. They appear to be covered with a layer of some substance which resists the stain, and which also enables them to resist various external agencies. This protective covering, together with their small amount of water, enables them to resist almost any amount of drying, a high degree of heat, and many other adverse conditions. Commonly the spores break out of the rod, and the rod producing them dies, although sometimes the rod may continue its activity even after the spores have been produced.

Arthrogenous spores (?).—Certain species of bacteria do not produce spores as just described, but may give rise to bodies that are sometimes called arthrospores. These bodies are formed as short segments of rods. A long rod may sometimes break up into several short rounded elements, which are clear and appear to have a somewhat increased power of resisting adverse conditions. The same may happen among the spherical forms, which only in rare instances form endogenous spores. Among the spheres which form a chain of streptococci some may occasionally be slightly different from the rest. They are a little larger, and have been thought to have an increased resisting power like that of true spores (Fig. 13 b). It is quite doubtful, however, whether it is proper to regard these bodies as spores. There is no good evidence that they have any special resisting power to heat like endogenous spores, and bacteriologists in general are inclined to regard them simply as resting cells. The term arthrospores has been given to them to indicate that they are formed as joints or segments, and this term may be a convenient one to retain although the bodies in question are not true spores.

Still a different method of spore formation occurs in a few peculiar bacteria. In this case (Fig. 14) the protoplasm in the large thread breaks into many minute spherical bodies, which finally find exit. The spores thus formed may not be all alike, differences in size being noticed. This method of spore formation occurs only in a few special forms of bacteria.

The matter of spore formation serves as one of the points for distinguishing species. Some species do not form spores, at least under any of the conditions in which they have been studied. Others form them readily in almost any condition, and others again only under special conditions which are adverse to their life. The method of spore formation is always uniform for any single species. Whatever be the method of the formation of the spore, its purpose in the life of the bacterium is always the same. It serves as a means of keeping the species alive under conditions of adversity. Its power of resisting heat or drying enables it to live where the ordinary active forms would be speedily killed. Some of these spores are capable of resisting a heat of 180 degrees C. (360 degrees F.) for a short time, and boiling water they can resist for a long time. Such spores when subsequently placed under favourable conditions will germinate and start bacterial activity anew.

MOTION.

Some species of bacteria have the power of active motion, and may be seen darting rapidly to and fro in the liquid in which they are growing. This motion is produced by flagella which protrude from the body. These flagella (Fig. 15) arise from a membrane surrounding the bacterium, but have an intimate connection with the protoplasmic content. Their distribution is different in different species of bacteria. Some species have a single flagellum at one end (Fig. 15 a). Others have one at each end (Fig. 15 b). Others, again, have, at least just before dividing, a bunch at one or both ends (Fig. 15 c and d), while others, again, have many flagella distributed all over the body in dense profusion (Fig. 15 e). These flagella keep up a lashing to and fro in the liquid, and the lashing serves to propel the bacteria through the liquid.

INTERNAL STRUCTURE.

It is hardly possible to say much about the structure of the bacteria beyond the description of their external forms. With all the variations in detail mentioned, they are extraordinarily simple, and about all that can be seen is their external shape. Of course, they have some internal structure, but we know very little in regard to it. Some microscopists have described certain appearances which they think indicate internal structure. Fig. 16 shows some of these appearances. The matter is as yet very obscure, however. The bacteria appear to have a membranous covering which sometimes is of a cellulose nature. Within it is protoplasm which shows various uncertain appearances. Some microscopists have thought they could find a nucleus, and have regarded bacteria as cells with inclosed nucleii (Figs. 10 a and 15 f). Others have regarded the whole bacterium as a nucleus without any protoplasm, while others, again, have concluded that the discerned internal structure is nothing except an appearance presented by the physical arrangement of the protoplasm. While we may believe that they have some internal structure, we must recognise that as yet microscopists have not been able to make it out. In short, the bacteria after two centuries of study appear to us about as they did at first. They must still be described as minute spheres, rods, or spirals, with no further discernible structure, sometimes motile and sometimes stationary, sometimes producing spores and sometimes not, and multiplying universally by binary fission. With all the development of the modern microscope we can hardly say more than this. Our advance in knowledge of bacteria is connected almost wholly with their methods of growth and the effects they produce in Nature.

ANIMALS OR PLANTS?

There has been in the past not a little question as to whether bacteria should be rightly classed with plants or with animals. They certainly have characters which ally them with both. Their very common power of active independent motion and their common habit of living upon complex bodies for foods are animal characters, and have lent force to the suggestion that they are true animals. But their general form, their method of growth and formation of threads, and their method of spore formation are quite plantlike. Their general form is very similar to a group of low green plants known as Oscillaria. Fig. 17 shows a group of these Oscillariae, and the similarity of this to some of the thread-like bacteria is decided. The Oscillariae are, however, true plants, and are of a green colour. Bacteria are therefore to- day looked upon as a low type of plant which has no chlorophyll, [Footnote: Chlorophyll is the green colouring matter of plants.] but is related to Oscillariae. The absence of the chlorophyll has forced them to adopt new relations to food, and compels them to feed upon complex foods instead of the simple ones, which form the food of green plants. We may have no hesitation, then, in calling them plants. It is interesting to notice that with this idea their place in the organic world is reduced to a small one systematically. They do not form a class by themselves, but are simply a subclass, or even a family, and a family closely related to several other common plants. But the absence of chlorophyll and the resulting peculiar life has brought about a curious anomaly. Whereas their closest allies are known only to botanists, and are of no interest outside of their systematic relations, the bacteria are familiar to every one, and are demanding the life attention of hundreds of investigators. It is their absence of chlorophyll and their consequent dependence upon complex foods which has produced this anomaly.

CLASSIFICATION OF BACTERIA.

While it has generally been recognised that bacteria are plants, any further classification has proved a matter of great difficulty, and bacteriologists find it extremely difficult to devise means of distinguishing species. Their extreme simplicity makes it no easy matter to find points by which any species can be recognised. But in spite of their similarity, there is no doubt that many different species exist. Bacteria which appear to be almost identical, under the microscope prove to have entirely different properties, and must therefore be regarded as distinct species. But how to distinguish them has been a puzzle. Microscopists have come to look upon the differences in shape, multiplication, and formation of spores as furnishing data sufficient to enable them to divide the bacteria into genera. The genus Bacillus, for instance, is the name given to all rod-shaped bacteria which form endogenous spores, etc. But to distinguish smaller subdivisions it has been found necessary to fall back upon other characters, such as the shape of the colony produced in solid gelatine, the power to produce disease, or to oxidize nitrites, etc. Thus at present the different species are distinguished rather by their physiological than their morphological characters. This is an unsatisfactory basis of classification, and has produced much confusion in the attempts to classify bacteria. The problem of determining the species of bacteria is to-day a very difficult one, and with our best methods is still unsatisfactorily solved. A few species of marked character are well known, and their powers of action so well understood that they can be readily recognised; but of the great host of bacteria studied, the large majority have been so slightly experimented upon that their characters are not known, and it is impossible, therefore, to distinguish many of them apart. We find that each bacteriologist working in any special line commonly keeps a list of the bacteria which he finds, with such data in regard to them as he has collected. Such a list is of value to him, but commonly of little value to other bacteriologists from the insufficiency of the data. Thus it happens that a large part of the different species of bacteria described in literature to- day have been found and studied by one investigator alone. By him they have been described and perhaps named. Quite likely the same species may have been found by two or three other bacteriologists, but owing to the difficulty of comparing results and the incompleteness of the descriptions the identity of the species is not discovered, and they are probably described again under different names. The same process may be repeated over and over again, until the same species of bacterium will come to be known by several different names, as it has been studied by different observers.

VARIATION OF BACTERIA.

This matter is made even more confusing by the fact that any species of bacterium may show more or less variation. At one time in the history of bacteriology, a period lasting for many years, it was the prevalent opinion that there was no constancy among bacteria, but that the same species might assume almost any of the various forms and shapes, and possess various properties. Bacteria were regarded by some as stages in the life history of higher plants. This question as to whether bacteria remain constant in character for any considerable length of time has ever been a prominent one with bacteriologists, and even to-day we hardly know what the final answer will be. It has been demonstrated beyond peradventure that some species may change their physiological characters. Disease bacteria, for instance, under certain conditions lose their powers of developing disease. Species which sour milk, or others which turn gelatine green, may lose their characters. Now, since it is upon just such physiological characters as these that we must depend in order to separate different species of bacteria from each other, it will be seen that great confusion and uncertainty will result in our attempts to define species. Further, it has been proved that there is sometimes more or less of a metamorphosis in the life history of certain species of bacteria. The same species may form a short rod, or a long thread, or break up into spherical spores, and thus either a short rod, or a thread, or a spherical form may belong to the same species. Other species may be motile at one time and stationary at another, while at a third period it is a simple mass of spherical spores. A spherical form, when it lengthens before dividing, appears as a short rod, and a short rod form after dividing may be so short as to appear like a spherical organism.

With all these reasons for confusion, it is not to be wondered at that no satisfactory classification of bacteria has been reached, or that different bacteriologists do not agree as to what constitutes a species, or whether two forms are identical or not. But with all the confusion there is slowly being obtained something like system. In spite of the fact that species may vary and show different properties under different conditions, the fundamental constancy of species is everywhere recognised to-day as a fact. The members of the same species may show different properties under different conditions, but it is believed that under identical conditions the properties will be constant. It is no more possible to convert one species into another than it is among the higher orders of plants. It is believed that bacteria do form a group of plants by themselves, and are not to be regarded as stages in the history of higher plants. It is believed that, together with a considerable amount of variability and an occasional somewhat long life history with successive stages, there is also an essential constancy. A systematic classification has been made which is becoming more or less satisfactory. We are constantly learning more and more of the characters, so that they can be recognised in different places by different observers. It is the conviction of all who work with bacteria that, in spite of the difficulties, it is only a matter of time when we shall have a classification and description of bacteria so complete as to characterize the different species accurately.

Even with our present incomplete knowledge of what characterizes a species, it is necessary to use some names. Bacteria are commonly given a generic name based upon their microscopic appearance. There are only a few of these names. Micrococcus, Streptococcus, Staphylococcus, Sarcina, Bacterium, Bacillus, Spirillum, are all the names in common use applying to the ordinary bacteria. There are a few others less commonly used. To this generic name a specific name is commonly added, based upon some physiological character. For example, Bacillus typhosus is the name given to the bacillus which causes typhoid fever. Such names are of great use when the species is a common and well-known one, but of doubtful value for less-known species It frequently happens that a bacteriologist makes a study of the bacteria found in a certain locality, and obtains thus a long list of species hitherto unknown. In these cases it is common simply to number these species rather than name them. This method is frequently advisable, since the bacteriologist can seldom hunt up all bacteriological literature with sufficient accuracy to determine whether some other bacteriologist may not have found the same species in an entirely different locality. One bacteriologist, for example, finds some seventy different species of bacteria in different cheeses. He studies them enough for his own purposes, but not sufficiently to determine whether some other person may not have found the same species perhaps in milk or water. He therefore simply numbers them—a method which conveys no suggestion as to whether they may be new species or not. This method avoids the giving of separate names to the same species found by different observers, and it is hoped that gradually accumulating knowledge will in time group together the forms which are really identical, but which have been described by different observers.

WHERE BACTERIA ARE FOUND.

There are no other plants or animals so universally found in Nature as the bacteria. It is this universal presence, together with their great powers of multiplication, which renders them of so much importance in Nature. They exist almost everywhere on the surface of the earth. They are in the soil, especially at its surface. They do not extend to very great depths of soil, however, few existing below four feet of soil. At the surface they are very abundant, especially if the soil is moist and full of organic material. The number may range from a few hundred to one hundred millions per gramme. [Footnote: One gramme is fifteen grains.] The soil bacteria vary also in species, some two-score different species having been described as common in soil. They are in all bodies of water, both at the surface and below it. They are found at considerable depths in the ocean. All bodies of fresh water contain them, and all sediments in such bodies of water are filled with bacteria. They are in streams of running water in even greater quantity than in standing water. This is simply because running streams are being constantly supplied with water which has been washing the surface of the country and thus carrying off all surface accumulations. Lakes or reservoirs, however, by standing quiet allow the bacteria to settle to the bottom, and the water thus gets somewhat purified. They are in the air, especially in regions of habitation. Their numbers are greatest near the surface of the ground, and decrease in the upper strata of air. Anything which tends to raise dust increases the number of bacteria in the air greatly, and the dust and emanations from the clothes of people crowded in a close room fill the air with bacteria in very great numbers. They are found in excessive abundance in every bit of decaying matter wherever it may be. Manure heaps, dead bodies of animals, decaying trees, filth and slime and muck everywhere are filled with them, for it is in such places that they find their best nourishment. The bodies of animals contain them in the mouth, stomach, and intestine in great numbers, and this is, of course, equally true of man. On the surface of the body they cling in great quantity; attached to the clothes, under the finger nails, among the hairs, in every possible crevice or hiding place in the skin, and in all secretions. They do not, however, occur in the tissues of a healthy individual, either in the blood, muscle, gland, or any other organ. Secretions, such as milk, urine, etc., always contain them, however, since the bacteria do exist in the ducts of the glands which conduct the secretions to the exterior, and thus, while the bacteria are never in the healthy gland itself, they always succeed in contaminating the secretion as it passes to the exterior. Not only higher animals, but the lower animals also have their bodies more or less covered with bacteria. Flies have them on their feet, bees among their hairs, etc.

In short, wherever on the face of Nature there is a lodging place for dust there will be found bacteria. In most of these localities they are dormant, or at least growing only a little. The bacteria clinging to the dry hair can grow but little, if at all, and those in pure water multiply very little. When dried as dust they are entirely dormant.

But each individual bacterium or spore has the potential power of multiplication already noticed, and as soon as it by accident falls upon a place where there is food and moisture it will begin to multiply. Everywhere in Nature, then, exists this group of organisms with its almost inconceivable power of multiplication, but a power held in check by lack of food. Furnish them with food and their potential powers become actual. Such food is provided by the dead bodies of animals or plants, or by animal secretions, or from various other sources. The bacteria which are fortunate enough to get furnished with such food material continue to feed upon it until the food supply is exhausted or their growth is checked in some other way. They may be regarded, therefore, as a constant and universal power usually held in check. With their universal presence and their powers of producing chemical changes in food material, they are ever ready to produce changes in the face of Nature, and to these changes we will now turn.

CHAPTER II.

MISCELLANEOUS USE OF BACTERIA IN THE ARTS.

The foods upon which bacteria live are in endless variety, almost every product of animal or vegetable life serving to supply their needs. Some species appear to require somewhat definite kinds of food, and have therefore rather narrow conditions of life, but the majority may live upon a great variety of organic compounds. As they consume the material which serves them as food they produce chemical changes therein. These changes are largely of a nature that the chemist knows as decomposition changes. By this is meant that the bacteria, seizing hold of ingredients which constitute their food, break them to pieces chemically. The molecule of the original food matter is split into simpler molecules, and the food is thus changed in its chemical nature. As a result, the compounds which appear in the decomposing solution are commonly simpler than the original food molecules. Such products are in general called decomposition products, or sometimes cleavage products. Sometimes, however, the bacteria have, in addition to their power of pulling their food to pieces, a further power of building other compounds out of the fragments, thus building up as well as pulling down. But, however they do it, bacteria when growing in any food material have the power of giving rise to numerous products which did not exist in the food mass before. Because of their extraordinary powers of reproduction they are capable of producing these changes very rapidly and can give rise in a short time to large amounts of the peculiar products of their growth.

It is to these powers of producing chemical changes in their food that bacteria owe all their importance in the world. Their power of chemically destroying the food products is in itself of no little importance, but the products which arise as the result of this series of chemical changes are of an importance in the world which we are only just beginning to appreciate. In our attempt to outline the agency which bacteria play in our industries and in natural processes as well, we shall notice that they are sometimes of value simply for their power of producing decomposition; but their greatest value lies in the fact that they are important agents because of the products of their life.

We may notice, in the first place, that in the arts there are several industries which may properly be classed together as maceration industries, all of which are based upon the decomposition powers of bacteria. Hardly any animal or vegetable substance is able to resist their softening influence, and the artisan relies upon this power in several different directions.

BENEFITS DERIVED FROM POWERS OF DECOMPOSITION.

Linen.—Linen consists of certain woody fibres of the stem of the flax. The flax stem is not made up entirely of the valuable fibres, but largely of more brittle wood fibres, which are of no use. The valuable fibres are, however, closely united with the wood and with each other in such an intimate fashion that it is impossible to separate them by any mechanical means. The whole cellular substance of the stem is bound together by some cementing materials which hold it in a compact mass, probably a salt of calcium and pectinic acid. The art of preparing flax is a process of getting rid of the worthless wood fibres and preserving the valuable, longer, tougher, and more valuable fibres, which are then made into linen. But to separate them it is necessary first to soften the whole tissue. This is always done through the aid of bacteria. The flax stems, after proper preparation, are exposed to the action of moisture and heat, which soon develops a rapid bacterial growth. Sometimes this is done by simply exposing the flax to the dew and rain and allowing it to lie thus exposed for some time. By another process the stems are completely immersed in water and allowed to remain for ten to fourteen days. By a third process the water in which the flax is immersed is heated from 75 degrees to 90 degrees F., with the addition of certain chemicals, for some fifty to sixty hours. In all cases the effect is the same. The moisture and the heat cause a growth of bacteria which proceeds with more or less rapidity according to the temperature and other conditions. A putrefactive fermentation is thus set up which softens the gummy substance holding the fibres together. The process is known as "retting," and after it is completed the fibres are easily isolated from each other. A purely mechanical process now easily separates the valuable fibres from the wood fibres. The whole process is a typical fermentation. A disagreeable odour arises from the fermenting flax, and the liquid after the fermentation is filled with products which make valuable manure. The process has not been scientifically studied until very recently. The bacillus which produces the "retting" is known now, however, and it has been shown that the "retting" is a process of decomposition of the pectin cement. No method of separating the linen fibres in the flax from the wood fibres has yet been devised which dispenses with the aid of bacteria.

Jute and Hemp.—Almost exactly the same use is made of bacterial action in the manufacture of jute und hemp. The commercial aspect of the jute industry has grown to be a large one, involving many millions of dollars. Like linen, jute is a fibre of the inner bark of a plant, and is mixed in the bark with a mass of other useless fibrous material. As in the case of linen, a fermentation by bacteria is depended upon as a means of softening the material so that the fibres can be disassociated. The process is called "retting," as in the linen manufacture. The details of the process are somewhat different. The jute is commonly fermented in tanks of stagnant water, although sometimes it is allowed to soak in river water for a sufficient length of time to produce the softening. After the fermentation is thus started the jute fibre is separated from the wood, and is of a sufficient flexibility and toughness to be woven into sacking, carpets, curtains, table covers, and other coarse cloth.

Practically the same method is used in separating the tough fibres of the hemp. The hemp plant contains some long flexible fibres with others of no value, and bacterial fermentation is relied upon to soften the tissues so that they may be separated.

Cocoanut fibre, a somewhat similar material is obtained from the husk of the cocoanut by the same means. The unripened husk is allowed to steep and ferment in water for a long time, six months or a year being required. By this time the husk has become so softened that it can be beaten until the fibres separate and can be removed.

They are subsequently made into a number of coarse articles, especially valuable for their toughness. Door mats, brushes, ships' fenders, etc., are illustrations of the sort of articles made from them.

In each of these processes the fermentation must have a tendency to soften the desired fibres as well as the connecting substance. Putrefaction attacks all kinds of vegetable tissue, and if this "retting" continues too long the desired fibre is decidedly injured by the softening effect of the fermentation. It is quite probable that, even as commonly carried on, the fermentation has some slight injurious effect upon the fibre, and that if some purely mechanical means could be devised for separating the fibre from the wood it would produce a better material. But such mechanical means has not been devised, and at present a putrefactive fermentation appears to be the only practical method of separating the fibres.

Sponges. — A somewhat similar use is made of bacteria in the commercial preparation of sponges. The sponge of commerce is simply the fibrous skeleton of a marine animal. When it is alive this skeleton is completely filled with the softer parts of the animal, and to fit the sponge for use this softer organic material must be got rid of. It is easily accomplished by rotting. The fresh sponges are allowed to stand in the warm sun and very rapidly decay. Bacteria make their way into the sponge and thoroughly decompose the soft tissues. After a short putrefaction of this sort the softened organic matter can be easily washed out of the skeleton and leave the clean fibre ready for market.

Leather preparation. — The tanning of leather is a purely chemical process, and in some processes the whole operation of preparing the leather is a chemical one. In others, however, especially in America, bacteria are brought into action at one stage. The dried hide which comes to the tannery must first have the hair removed together with the outer skin. The hide for this purpose must be moistened and softened. In some tanneries this is done by steeping it in chemicals. In others, however, it is put into water and slightly heated until fermentation arises. The fermentation softens it so that the outer skin can be easily removed with a knife, and the removal of hair is accomplished at the same time. Bacterial putrefaction in the tannery is thus an assistance in preparing the skin for the tanning proper.

Even in the subsequent tanning a bacterial fermentation appears to play a part, but little is yet known in regard to it.

Maceration of skeletons. — The making of skeletons for museums and anatomical instruction in general is no very great industry, and yet it is one of importance. In the making of skeletons the process of maceration is commonly used as an aid. The maceration consists simply in allowing the skeleton to soak in water for a day or two after cleaning away the bulk of the muscles. The putrefaction that arises softens the connective tissues so much that the bones may be readily cleaned of flesh.

Citric acid. — Bacterial fermentation is employed also in the ordinary preparation of citric acid. The acid is made chiefly from the juice of the lemon. The juice is pressed from the fruit and then allowed to ferment. The fermentation aids in separating a mucilaginous mass and making it thus possible to obtain the citric acid in a purer condition. The action is probably similar to the maceration processes described above, although it has not as yet been studied by bacteriologists.

BENEFITS DERIVED FROM THE PRODUCTS OF BACTERIAL LIFE.

While bacteria thus play a part in our industries simply from their power of producing decomposition, it is primarily because of the products of their action that they are of value. Wherever bacteria seize hold of organic matter and feed upon it, there are certain to be developed new chemical compounds, resulting largely from decomposition, but partly also from constructive processes. These new compounds are of great variety. Different species of bacteria do not by any means produce the same compounds even when growing in and decomposing the same food material. Moreover, the same species of bacteria may give rise to different products when growing in different food materials. Some of the compounds produced by such processes are poisonous, others are harmless. Some are gaseous, others are liquids. Some have peculiar odours, as may be recognised from the smell arising from a bit of decaying meat. Others have peculiar tastes, as may be realized in the gamy taste of meat which is in the incipient stages of putrefaction. By purely empirical means mankind has learned methods of encouraging the development of some of these products, and is to-day making practical use of this power, possessed by bacteria, of furnishing desired chemical compounds. Industries involving the investment of hundreds of millions of dollars are founded upon the products of bacterial life, and they have a far more important relation to our everyday life than is commonly imagined. In many cases the artisan who is dependent upon this action of microscopic life is unaware of the fact. His processes are those which experience has taught produce desired results, but, nevertheless, his dependence upon bacteria is none the less fundamental.

BACTERIA IN THE FERMENTATIVE INDUSTRIES.

We may notice, first, several miscellaneous instances of the application of bacteria to various fermentative industries where their aid is of more or less value to man. In some of the examples to be mentioned the influence of bacteria is profound and fundamental, while in others it is only incidental. The fermentative industries of civilization are gigantic in extent, and have come to be an important factor in modern civilized life. The large part of the fermentation is based upon the growth of a class of microscopic plants which we call yeasts. Bacteria and yeasts are both microscopic plants, and perhaps somewhat closely related to each other. The botanist finds a difference between them, based upon their method of multiplication, and therefore places them in different classes (Fig. 2, page 19). In their general power of producing chemical changes in their food products, yeasts agree closely with bacteria, though the kinds of chemical changes are different. The whole of the great fermentative industries, in which are invested hundreds of millions of dollars, is based upon chemical decompositions produced by microscopic plants. In the great part of commercial fermentations alcohol is the product desired, and alcohol, though it is sometimes produced by bacteria, is in commercial quantities produced only by yeasts. Hence it is that, although the fermentations produced by bacteria are more common in Nature than those produced by yeasts and give rise to a much larger number of decomposition products, still their commercial aspect is decidedly less important than that of yeasts. Nevertheless, bacteria are not without their importance in the ordinary fermentative processes. Although they are of no importance as aids in the common fermentative processes, they are not infrequently the cause of much trouble. In the fermentation of malt to produce beer, or grape juice to produce wine, it is the desire of the brewer and vintner to have this fermentation produced by pure yeasts, unmixed with bacteria. If the yeast is pure the fermentation is uniform and successful. But the brewer and vintner have long known that the fermentation is frequently interfered with by irregularities. The troubles which arise have long been known, but the bacteriologist has finally discovered their cause, and in general their remedy. The cause of the chief troubles which arise in the fermentation is the presence of contaminating bacteria among the yeasts. These bacteria have been more or less carefully studied by bacteriologists, and their effect upon the beer or wine determined. Some of them produce acid and render the products sour; others make them bitter; others, again, produce a slimy material which makes the wine or beer "ropy." Something like a score of bacteria species have been found liable to occur in the fermenting material and destroy the value of the product of both the wine maker and the beer brewer. The species of bacteria which infect and injure wine are different from those which infect and injure beer. They are ever present as possibilities in the great alcoholic fermentations. They are dangers which must be guarded against. In former years the troubles from these sources were much greater than they are at present. Since it has been demonstrated that the different imperfections in the fermentative process are due to bacterial impurities, commonly in the yeasts which are used to produce the fermentation, methods of avoiding them are readily devised. To-day the vintner has ready command of processes for avoiding the troubles which arise from bacteria, and the brewer is always provided with a microscope to show him the presence or absence of the contaminating bacteria. While, then, the alcoholic fermentations are not dependent upon bacteria, the proper management of these fermentations requires a knowledge of their habits and characters.

There are certain other fermentative processes of more or less importance in their commercial aspects, which are directly dependent upon bacterial action, Some of them we should unhesitatingly look upon as fermentations, while others would hardly be thought of as belonging to the fermentation industries.

VINEGAR.

The commercial importance of the manufacture of vinegar, though large, does not, of course, compare in extent with that of the alcoholic fermentations. Vinegar is a weak solution of acetic acid, together with various other ingredients which have come from the materials furnishing the acid. In the manufacture of vinegar, alcohol is always used as the source of the acetic acid. The production of acetic acid from alcohol is a simple oxidation. The equation $C_2H_6O + O_2 = C_2H_4O_2 + H_2O$ shows the chemical change that occurs. This oxidation can be brought about by purely chemical means. While alcohol will not readily unite with oxygen under common conditions, if the alcohol is allowed to pass over a bit of platinum sponge the union readily occurs and acetic acid results. This method of acetic-acid production is possible experimentally, but is impracticable on any large scale. In the ordinary manufacture of vinegar the oxidation is a true fermentation, and brought about by the growth of bacteria.

In the commercial manufacture of vinegar several different weak alcoholic solutions are used. The most common of these are fermented malt, weak wine, cider, and sometimes a weak solution of spirit to which is added sugar and malt. If these solutions are allowed to stand for a time in contact with air, they slowly turn sour by the gradual conversion of the alcohol into acetic acid. At the close of the process practically all of the alcohol has disappeared. Ordinarily, however, not all of it has been converted into acetic acid, for the oxidation does not all stop at this step. As the oxidation goes on, some of the acid is oxidized into carbonic dioxide, which is, of course, dissipated at once into the air, and if the process is allowed to continue unchecked for a long enough period much of the acetic acid will be lost in this way.

The oxidation of the alcohol in all commercial production of vinegar is brought about by the growth of bacteria in the liquid. When the vinegar production is going on properly, there is formed on the top of the liquid a dense felted mass known as the "mother of vinegar." This mass proves to be made of bacteria which have the power of absorbing oxygen from the air, or, at all events, of causing the alcohol to unite with oxygen. It was at first thought that a single species of bacterium was thus the cause of the oxidation of alcohol, and this was named Mycoderma aceti. But further study has shown that several have the power, and that even in the commercial manufacture of vinegar several species play a part (Fig. 18), although the different species are not yet very thoroughly studied. Each appears to act best under different conditions. Some of them act slowly, and others rapidly, the slow- growing species appearing to produce the larger amount of acid in the end. After the amount of acetic acid reaches a certain percentage, the bacteria are unable to produce more, even though there be alcohol still left unoxidized. A percentage as high as fourteen per cent, commonly destroys all their power of growth. The production of the acid is wholly dependent upon the growth of the bacteria, and the secret of the successful vinegar manufacture is the skilful manipulation of these bacteria so as to keep them in the purest condition and to give them the best opportunity for growth.

One method of vinegar manufacture which is quite rapid is carried on in a slightly different manner. A tall cylindrical chamber is filled with wood shavings, and a weak solution of alcohol is allowed to trickle slowly through it. The liquid after passing over the shavings comes out after a number of hours well charged with acetic acid. This process at first sight appears to be a purely chemical one, and reminds us of the oxidation which occurs when alcohol is allowed to pass over a platinum sponge. It has been claimed, indeed, that this is a chemical oxidation in which bacteria play no part. But this appears to be an error. It is always found necessary in this method to start the process by pouring upon the shavings some warm vinegar. Unless in this way the shavings become charged with the vinegar-holding bacteria the alcohol will not undergo oxidation during its passage over them, and after the bacteria thus introduced have grown enough to coat the shavings thoroughly the acetic-acid production is much more rapid than at first. If vinegar is allowed to trickle slowly down a suspended string, so that its bacteria may distribute themselves through the string, and then alcohol be allowed to trickle over it in the same way, the oxidation takes place and acetic acid is formed. From the accumulation of such facts it has come to be recognised that all processes for the commercial manufacture of vinegar depend upon the action of bacteria. While the oxidation of alcohol into acetic acid may take place by purely chemical means, these processes are not practical on a large scale, and vinegar manufacturers everywhere depend upon bacteria as their agents in producing the oxidation. These bacteria, several species in all, feed upon the nitrogenous matter in the fermenting mass and produce the desired change in the alcohol.

This vinegar fermentation is subject to certain irregularities, and the vinegar manufacturers can not always depend upon its occurring in a satisfactory manner. Just as in brewing, so here, contaminating bacteria sometimes find their way into the fermenting mass and interfere with its normal course. In particular, the flavour of the vinegar is liable to suffer from such causes. As yet our vinegar manufacturers have not applied to acetic fermentation the same principle which has been so successful in brewing—namely, the use, as a starter of the fermentation, of a pure culture of the proper species of bacteria. This has been done experimentally and proves to be feasible. In practice, however, vinegar makers find that simpler methods of obtaining a starter—by means of which they procure a culture nearly though not absolutely pure—are perfectly satisfactory. It is uncertain whether really pure cultures will ever be used in this industry.

LACTIC ACID.

The manufacture of lactic acid is an industry of less extent than that of acetic acid, and yet it is one which has some considerable commercial importance. Lactic acid is used in no large quantity, although it is of some value as a medicine and in the arts. For its production we are wholly dependent upon bacteria. It is this acid which, as we shall see, is produced in the ordinary souring of milk, and a large number of species of bacteria are capable of producing the acid from milk sugar. Any sample of sour milk may therefore always be depended upon to contain plenty of lactic organisms. In its manufacture for commercial purposes milk is sometimes used as a source, but more commonly other substances. Sometimes a mixture of cane sugar and tartaric acid is used. To start the fermentation the mixture is inoculated with a mass of sour milk or decaying cheese, or both, such a mixture always containing lactic organisms. To be sure, it also contains many other bacteria which have different effects, but the acid producers are always so abundant and grow so vigorously that the lactic fermentation occurs in spite of all other bacteria. Here also there is a possibility of an improvement in the process by the use of pure cultures of lactic organisms. Up to the present, however, there has been no application of such methods. The commercial aspects of the industry are not upon a sufficiently large scale to call for much in this direction.

At the present time the only method we have for the manufacture of lactic acid is dependent upon bacteria. Chemical processes for its manufacture are known, but not employed commercially. There are several different kinds of lactic acid. They differ from each other in the relations of the atoms within their molecule, and in their relation to polarized light, some forms rotating the plane of polarized light to the right, others to the left, while others are inactive in this respect. All the types are produced by fermentation processes, different species of bacteria having powers of producing the different types.

BUTYRIC ACID.

Butyric acid is another acid for which we are chiefly dependent upon bacteria. This acid is of no very great importance, and its manufacture can hardly be called an industry; still it is to a certain extent made, and is an article of commerce. It is an acid that can be manufactured by chemical means, but, as in the case of the last two acids, its commercial manufacture is based upon bacterial action. Quite a number of species of bacteria can produce butyric acid, and they produce it from a variety of different sources. Butyric acid is a common ingredient in old milk and in butter, and its formation by bacteria was historically one of the first bacterial fermentations to be clearly understood. It can be produced also in various sugar and starchy solutions. Glycerine may also undergo a butyric fermentation. The presence of this acid is occasionally troublesome, since it is one of the factors in the rancidity of butter and other similar materials.

INDIGO PREPARATION.

The preparation of indigo from the indigo plant is a fermentative process brought about by a specific bacterium. The leaves of the plant are immersed in water in a large vat, and a rapid fermentation arises. As a result of the fermentation the part of the plant which is the basis of the indigo is separated from the leaves and dissolved in the water; and as a second feature of the fermentation the soluble material is changed in its chemical nature into indigo proper. As this change occurs the characteristic blue colour is developed, and the material is rendered insoluble in water. It therefore makes its appearance as a blue mass separated from the water, and is then removed as indigo.

Of the nature of the process we as yet know very little. That it is a fermentation is certain, and it has been proved that it is produced by a definite species of bacterium which occurs on the indigo leaves. If the sterilized leaves are placed in sterile water no fermentation occurs and no indigo is formed. If, however, some of the specific bacteria are added to the mass the fermentation soon begins and the blue colour of the indigo makes its appearance. It is plain, therefore, that indigo is a product of bacterial fermentation, and commonly due to a single definite species of bacterium. Of the details of the formation, however, we as yet know little, and no practical application of the facts have yet been made.

BACTERIA IN TOBACCO CURING.

A fermentative process of quite a different nature, but of immense commercial value, is found in the preparation of tobacco. The process by which tobacco is prepared is a long and somewhat complicated one, consisting of a number of different stages. The tobacco, after being first dried in a careful manner, is subsequently allowed to absorb moisture from the atmosphere, and is then placed in large heaps to undergo a further change. This process appears to be a fermentation, for the temperature of the mass rises rapidly, and every indication of a fermentative action is seen. The tobacco in these heaps is changed occasionally, the heap being thrown down and built up again in such a way that the portion which was first at the bottom comes to the top, and in this way all parts of the heap may become equally affected by the process. After this process the tobacco is sent to the different manufacturers, who finish the process of curing. The further treatment it receives varies widely according to the desired product, whether for smoking or for snuff, etc. In all cases, however, fermentations play a prominent part. Sometimes the leaves are directly inoculated with fermenting material. In the preparation of snuff the details of the process are more complicated than in the preparation of smoking tobacco. The tobacco, after being ground and mixed with certain ingredients, is allowed to undergo a fermentation which lasts for weeks, and indeed for months. In the different methods of preparing snuff the fermentations take place in different ways, and sometimes the tobacco is subjected to two or three different fermentative actions. The result of the whole is the slow preparation of the commercial product. It is during the final fermentative processes that the peculiar colour and flavour of the snuff are developed, and it is during the fermentation of the leaves of the smoking tobacco—either the original fermentation or the subsequent ones— that the special flavours and aromas of tobacco are produced.

It can not be claimed for a moment that these changes by which the tobacco is cured and finally brought to a marketable condition are due wholly to bacteria. There is no question that chemical and physical phenomena play an important part in them. Nevertheless, from the moment when the tobacco is cut in the fields until the time it is ready for market the curing is very intimately associated with bacteria and fermentative organisms in general. Some of these processes are wholly brought about by bacterial life; in others the micro-organisms aid the process, though they perhaps can not be regarded as the sole agents.

At the outset the tobacco producer has to contend with a number of micro-organisms which may produce diseases in his tobacco. During the drying process, if the temperature or the amount of moisture or the access of air is not kept in a proper condition, various troubles arise and various diseases make their appearance, which either injure or ruin the value of the product. These appear to be produced by micro-organisms of different sorts. During the fermentation which follows the drying the producer has to contend with micro-organisms that are troublesome to him; for unless the phenomena are properly regulated the fermentation that occurs produces effects upon the tobacco which ruin its character. From the time the tobacco is cut until the final stage in the curing the persons engaged in preparing it for market must be on a constant watch to prevent the growth within it of undesirable organisms. The preparation of tobacco is for this reason a delicate operation, and one that will be very likely to fail unless the greatest care is taken. In the several fermentative processes which occur in the preparation there is no question that micro-organisms aid the tobacco producer and manufacturer. Bacteria produce the first fermentation that follows the drying, and it is these organisms too, in large measure, that give rise to all the subsequent fermentations, although seemingly in some cases purely chemical processes materially aid. Now the special quality of the tobacco is in part dependent upon the peculiar type of fermentation which occurs in one or another of these fermenting actions. It is the fermentation that gives rise to the peculiar flavour and to the aroma of the different grades of tobacco. Inasmuch as the various flavours which characterize tobacco of different grades are developed, at least to a large extent, during the fermentation processes, it is a natural supposition that the different qualities of the tobacco, so far as concerns flavour, are due to the different types of fermentation. The number of species of bacteria which are found upon the tobacco leaves in the various stages of its preparation is quite large, and from what we have already learned it is inevitable that the different kinds of bacteria will produce different results in the fermenting process. It would seem natural, therefore, to assume that the different flavours of different grades may not unlikely be due to the fact that the tobacco in the different cases has been fermented under the influence of different kinds of bacteria.

Nor is this simply a matter of inference. To a certain extent experimental evidence has borne out the conclusion, and has given at least a slight indication of practical results in the future. Acting upon the suggestion that the difference between the high grades of tobacco and the poorer grades is due to the character of the bacteria that produce the fermentation, certain bacteriologists have attempted to obtain from a high quality of tobacco the species of bacteria which are infesting it. These bacteria have then been cultivated by bacteriological methods and used in experiments for the fermentation of tobacco. If it is true that the flavour of high grade tobacco is in large measure, or even in part, due to the action of the peculiar microbes from the soil where it grows, it ought to be possible to produce similar flavours in the leaves of tobacco grown in other localities, if the fermentation of the leaves is carried on by means of the pure cultures of bacteria obtained from the high grade tobacco. Not very much has been done or is known in this connection as yet. Two bacteriologists have experimented independently in fermenting tobacco leaves by the action of pure cultures of bacteria obtained from such sources. Each of them reports successful experiments. Each claims that they have been able to improve the quality of tobacco by inoculating the leaves with a pure culture of bacteria obtained from tobacco having high quality in flavour. In addition to this, several other bacteriologists have carried on experiments sufficient to indicate that the flavours of the tobacco and the character of the ripening may be decidedly changed by the use of different species of micro-organisms in the fermentations that go on during the curing processes.

In regard to the whole matter, however, we must recognise that as yet we have very little knowledge. The subject has been under investigation for only a short time; and, while considerable information has been derived, this information is not thoroughly understood, and our knowledge in regard to the matter is as yet in rather a chaotic condition. It seems certain, however, that the quality of tobacco is in large measure dependent upon the character of the fermentations that occur at different stages of the curing. It seems certain also that these fermentations are wholly or chiefly produced by microorganisms, and that the character of the fermentation is in large measure dependent upon the species of micro-organisms that produce it. If these are facts, it would seem not improbable that a further study may produce practical results for this great industry. The study of yeasts and the methods of keeping yeast from contaminations has revolutionised the brewing industry. Perhaps in this other fermentative industry, which is of such great commercial extent, the use of pure cultures of bacteria may in the future produce as great revolutions in methods as it has in the industry of the alcoholic fermentation.

It must not, however, be inferred that the differences in grades of tobacco grown in different parts of the world are due solely to variations in the curing processes and to the types of fermentation. There are differences in the texture of the leaves, differences in the chemical composition of the tobaccoes, which are due undoubtedly to the soils and the climatic conditions in which they grow, and these, of course, will never be affected by changing the character of the ferment active processes. It is, however, probable that in so far as the flavours that distinguish the high and low grades of tobacco are due to the character of the fermentative processes, they may be in the future, at least to a large extent, controlled by the use of pure cultures in curing processes. Seemingly, then, there is as great a future in the development of this fermentative industry as there has been in the past in the development of the fermentative industry associated with brewing and vinting.

OPIUM.

Opium for smoking purposes is commonly allowed to undergo a curing process which lasts several months. This appears to be somewhat similar to the curing of tobacco. Apparently it is a fermentation due to the growth of microorganisms. The organisms in question are not, however, bacteria in this case, but a species of allied fungus. The plant is a mould, and it is claimed that inoculation of the opium with cultures of this mould hastens the curing.

TROUBLESOME FERMENTATIONS.

Before leaving this branch of the subject it is necessary to notice some of the troublesome fermentations which are ever interfering with our industries, requiring special methods, or, indeed, sometimes developing special industries to meet them. As agents of decomposition, bacteria will of course be a trouble whenever they get into material which it is desired to preserve. Since they are abundant everywhere, it is necessary to count upon their attacking with certainty any fermentable substance which is exposed to air and water. Hence they are frequently the cause of much trouble. In the fermentative industries they occasionally cause an improper sort of fermentation to occur unless care is taken to prevent undesired species of bacteria from being present. In vinegar making, improper species of bacteria obtaining access to the solution give rise to undesirable flavours, greatly injuring the product. In tobacco curing it is very common for the wrong species of bacteria to gain access to the tobacco at some stage of the curing and by their growth give rise to various troubles. It is the ubiquitous presence of bacteria which makes it impossible to preserve fruits, meats, or vegetables for any length of time without special methods. This fact in itself has caused the development of one of our most important industries. Canning meats or fruits consists in nothing more than bringing them into a condition in which they will be preserved from attack of these micro-organisms. The method is extremely simple in theory. It is nothing more than heating the material to be preserved to a high temperature and then sealing it hermetically while it is still hot. The heat kills all the bacteria which may chance to be lodged in it, and the hermetical sealing prevents other bacteria from obtaining access. Inasmuch as all organic decomposition is produced by bacterial growth, such sterilized and sealed material will be preserved indefinitely when the operation is performed carefully enough. The methods of accomplishing this with sufficient care are somewhat varied in different industries, but they are all fundamentally the same. It is an interesting fact that this method of preserving meats was devised in the last century, before the relation of micro-organisms to fermentation and putrefaction was really suspected. For a long time it had been in practical use while scientists were still disputing whether putrefaction could be avoided by preventing the access of bacteria. The industry has, however, developed wonderfully within the last few years, since the principles underlying it have been understood. This understanding has led to better methods of destroying bacterial life and to proper sealing, and these have of course led to greater success in the preservation, until to-day the canning industries are among those which involve capital reckoned in the millions.

Occasionally bacteria are of some value in food products. The gamy flavour of meats is nothing more than incipient decomposition. Sauer Kraut is a food mass intentionally allowed to ferment and sour. The value of bacteria in producing butter and cheese flavours is noticed elsewhere. But commonly our aim must be to prevent the growth of bacteria in foods. Foods must be dried or cooked or kept on ice, or some other means adopted for preventing bacterial growth in them. It is their presence that forces us to keep our ice box, thus founding the ice business, as well as that of the manufacture of refrigerators. It is their presence, again, that forces us to smoke hams, to salt mackerel, to dry fish or other meats, to keep pork in brine, and to introduce numerous other details in the methods of food preparation and preservation.

CHAPTER III.

RELATION OF BACTERIA TO THE DAIRY INDUSTRY.

Dairying is one of the most primitive of our industries. From the very earliest period, ever since man began to keep domestic cattle, he has been familiar with dairying. During these many centuries certain methods of procedure have been developed which produce desired results. These methods, however, have been devised simply from the accumulation of experience, with very little knowledge as to the reasons underlying them. The methods of past centuries are, however, ceasing to be satisfactory. The advance of our civilization during the last half century has seen a marked expansion in the extent of the dairy industry. With this expansion has appeared the necessity for new methods, and dairymen have for years been looking for them. The last few years have been teaching us that the new methods are to be found along the line of the application of the discoveries of modern bacteriology. We have been learning that the dairyman is more closely related to bacteria and their activities than almost any other class of persons. Modern dairying, apart from the matter of keeping the cow, consists largely in trying to prevent bacteria from growing in milk or in stimulating their growth in cream, butter, and cheese. These chief products of the dairy will be considered separately.

SOURCES OF BACTERIA IN MILK.

The first fact that claims our attention is, that milk at the time it is secreted from the udder of the healthy cow contains no bacteria. Although bacteria are almost ubiquitous, they are not found in the circulating fluids of healthy animals, and are not secreted by their glands. Milk when first secreted by the milk gland is therefore free from bacteria. It has taken a long time to demonstrate this fact, but it has been finally satisfactorily proved. Secondly, it has been demonstrated that practically all of the normal changes which occur in milk after its secretion are caused by the growth of bacteria. This, too, was long denied, and for quite a number of years after putrefactions and fermentations were generally acknowledged to be caused by the growth of micro- organisms, the changes which occurred in milk were excepted from the rule. The uniformity with which milk will sour, and the difficulty, or seeming impossibility, of preventing this change, led to the belief that the souring of milk was a normal change characteristic of milk, just as clotting is characteristic of blood. This was, however, eventually disproved, and it was finally demonstrated that, beyond a few physical changes connected with evaporation and a slight oxidation of the fat, milk, if kept free from bacteria, will undergo no change. If bacteria are not present, it will remain sweet indefinitely.

But it is impossible to draw milk from the cow in such a manner that it will be free from bacteria except by the use of precautions absolutely impracticable in ordinary dairying. As milk is commonly drawn, it is sure to be contaminated by bacteria, and by the time it has entered the milk pail it contains frequently as many as half a million, or even a million, bacteria in every cubic inch of the milk. This seems almost incredible, but it has been demonstrated in many cases and is beyond question. Since these bacteria are not in the secreted milk, they must come from some external sources, and these sources are the following:

The first in importance is the cow herself; for while her milk when secreted is sterile, and while there are no bacteria in her blood, nevertheless the cow is the most prolific source of bacterial contamination. In the first place, the milk ducts are full of them. After each milking a little milk is always left in the duct, and this furnishes an ideal place for bacteria to grow. Some bacteria from the air or elsewhere are sure to get into these ducts after the milking, and they begin at once to multiply rapidly. By the next milking they become very abundant in the ducts, and the first milk drawn washes most of them at once into the milk pail, where they can continue their growth in the milk. Again, the exterior of the cow's body contains them in abundance. Every hair, every particle of dirt, every bit of dried manure, is a lurking place for millions of bacteria. The hind quarters of a cow are commonly in a condition of much filth, for the farmer rarely grooms his cow, and during the milking, by her movements, by the switching of her tail, and by the rubbing she gets from the milker, no inconsiderable amount of this dirt and filth is brushed off and falls into the milk pail The farmer understands this source of dirt and usually feels it necessary to strain the milk after the milking. But the straining it receives through a coarse cloth, while it will remove the coarser particles of dirt, has no effect upon the bacteria, for these pass through any strainer unimpeded. Again, the milk vessels themselves contain bacteria, for they are never washed absolutely clean. After the most thorough washing which the milk pail receives from the kitchen, there will always be left many bacteria clinging in the cracks of the tin or in the wood, ready to begin to grow as soon as the milk once more fills the pail The milker himself contributes to the supply, for he goes to the milking with unclean hands, unclean clothes, and not a few bacteria get from him to his milk pail. Lastly, we find the air of the milking stall furnishing its quota of milk bacteria. This source of bacteria is, however, not so great as was formerly believed. That the air may contain many bacteria in its dust is certain, and doubtless these fall in some quantity into the milk, especially if the cattle are allowed to feed upon dusty hay before and during the milking. But unless the air is thus full of dust this source of bacteria is not very great, and compared with the bacteria from the other sources the air bacteria are unimportant.

The milk thus gets filled with bacteria, and since it furnishes an excellent food these bacteria begin at once to grow. The milk when drawn is warm and at a temperature which especially stimulates bacterial growth. They multiply with great rapidity, and in the course of a few hours increase perhaps a thousandfold. The numbers which may be found after twenty-four hours are sometimes inconceivable; market milk may contain as many as five hundred millions per cubic inch; and while this is a decidedly extreme number, milk that is a day old will almost always contain many millions in each cubic inch, the number depending upon the age of the milk and its temperature. During this growth the bacteria have, of course, not been without their effect. Recognising as we do that bacteria are agents for chemical change, we are prepared to see the milk undergoing some modifications during this rapid multiplication of bacteria. The changes which these bacteria produce in the milk and its products are numerous, and decidedly affect its value. They are both advantageous and disadvantageous to the dairyman. They are nuisances so far as concerns the milk producer, but allies of the butter and cheese maker.

THE EFFECT OF BACTERIA ON MILK.

The first and most universal change effected in milk is its SOURING. So universal is this phenomenon that it is generally regarded as an inevitable change which can not be avoided, and, as already pointed out, has in the past been regarded as a normal property of milk. To-day, however, the phenomenon is well understood. It is due to the action of certain of the milk bacteria upon the milk sugar which converts it into lactic acid, and this acid gives the sour taste and curdles the milk. After this acid is produced in small quantity its presence proves deleterious to the growth of the bacteria, and further bacterial growth is checked. After souring, therefore, the milk for some time does not ordinarily undergo any further changes.

Milk souring has been commonly regarded as a single phenomenon, alike in all cases. When it was first studied by bacteriologists it was thought to be due in all cases to a single species of microorganism which was discovered to be commonly present and named Bacillus acidi lactici (Fig. 19). This bacterium has certainly the power of souring milk rapidly, and is found to be very common in dairies in Europe. As soon as bacteriologists turned their attention more closely to the subject it was found that the spontaneous souring of milk was not always caused by the same species of bacterium. Instead of finding this Bacillus acidi lactici always present, they found that quite a number of different species of bacteria have the power of souring milk, and are found in different specimens of soured milk. The number of species of bacteria which have been found to sour milk has increased until something over a hundred are known to have this power. These different species do not affect the milk in the same way. All produce some acid, but they differ in the kind and the amount of acid, and especially in the other changes which are effected at the same time that the milk is soured, so that the resulting soured milk is quite variable. In spite of this variety, however, the most recent work tends to show that the majority of cases of spontaneous souring of milk are produced by bacteria which, though somewhat variable, probably constitute a single species, and are identical with the Bacillus acidi lactici (Fig. 19). This species, found common in the dairies of Europe, according to recent investigations occurs in this country as well. We may say, then, that while there are many species of bacteria infesting the dairy which can sour the milk, there is one which is more common and more universally found than others, and this is the ordinary cause of milk souring.

When we study more carefully the effect upon the milk of the different species of bacteria found in the dairy, we find that there is a great variety of changes which they produce when they are allowed to grow in milk. The dairyman experiences many troubles with his milk. It sometimes curdles without becoming acid. Sometimes it becomes bitter, or acquires an unpleasant "tainted" taste, or, again, a "soapy" taste. Occasionally a dairyman finds his milk becoming slimy, instead of souring and curdling in the normal fashion. At such times, after a number of hours, the milk becomes so slimy that it can be drawn into long threads. Such an infection proves very troublesome, for many a time it persists in spite of all attempts made to remedy it. Again, in other cases the milk will turn blue, acquiring about the time it becomes sour a beautiful sky-blue colour. Or it may become red, or occasionally yellow. All of these troubles the dairyman owes to the presence in his milk of unusual species of bacteria which grow there abundantly.

Bacteriologists have been able to make out satisfactorily the connection of all these infections with different species of the bacteria. A large number of species have been found to curdle milk without rendering it acid, several render it bitter, and a number produce a "tainted" and one a "soapy" taste. A score or more have been found which have the power of rendering the milk slimy. Two different species at least have the power of turning the milk to sky-blue colour; two or three produce red pigments (Fig. 20), and one or two have been found which produce a yellow colour. In short, it has been determined beyond question that all these infections, which are more or less troublesome to dairymen, are due to the growth of unusual bacteria in the milk.

These various infections are all troublesome, and indeed it may be said that, so far as concerns the milk producer and the milk consumer, bacteria are from beginning to end a source of trouble. It is the desire of the milk producer to avoid them as far as possible — a desire which is shared also by everyone who has anything to do with milk as milk. Having recognised that the various troubles, which occasionally occur even in the better class of dairies, are due to bacteria, the dairyman is, at least in a measure, prepared to avoid them. The avoiding of these troubles is moderately easy as soon as dairymen recognise the source from which the infectious organisms come, and also the fact that low temperatures will in all cases remedy the evil to a large extent. With this knowledge in hand the avoidance of all these troubles is only a question of care in handling the dairy. It must be recognised that most of these troublesome bacteria come from some unusual sources of infection. By unusual sources are meant those which the exercise of care will avoid. It is true that the souring bacteria appear to be so universally distributed that they can not be avoided by any ordinary means. But all other troublesome bacteria appear to be within control. The milkman must remember that the sources of the troubles which are liable to arise in his milk are in some form of filth: either filth on the cow, or dust in the hay which is scattered through the barn, or dirt on cows' udders, or some other unusual and avoidable source. These sources, from what we have already noticed, will always furnish the milk with bacteria; but under common conditions, and when the cow is kept in conditions of ordinary cleanliness, and frequently even when not cleanly, will only furnish bacteria that produce the universal souring. Recognising this, the dairyman at once learns that his remedies for the troublesome infections are cleanliness and low temperatures. If he is careful to keep his milk vessels scrupulously clean; if he will keep his cow as cleanly as he does his horse; and if he will use care in and around the barn and dairy, and then apply low temperatures to the milk, he need never be disturbed by slimy or tainted milk, or any of these other troubles; or he can remove such infections speedily should they once appear. Pure sweet milk is only a question of sufficient care. But care means labour and expense. As long as we demand cheap milk, so long will we be supplied with milk procured under conditions of filth. But when we learn that cheap milk is poor milk, and when we are willing to pay a little more for it, then only may we expect the use of greater care in the handling of the milk, resulting in a purer product.

Bacteriology has therefore taught us that the whole question of the milk supply in our communities is one of avoiding the too rapid growth of bacteria. These organisms are uniformly a nuisance to the milkman. To avoid their evil influence have been designed all the methods of caring for the dairy and the barn, all the methods of distributing milk in ice cars. Moreover, all the special devices connected with the great industry of milk supply have for their foundation the attempt to avoid, in the first place, the presence of too great a number of bacteria, and. in the second place, the growth of these bacteria.

BACTERIA IN BUTTER MAKING.

CREAM RIPENING.—Passing from milk to butter, we find a somewhat different story, inasmuch as here bacteria are direct allies to the dairyman rather than his enemies. Without being aware of it, butter makers have for years been making use of bacteria in their butter making and have been profiting by the products which the bacteria have furnished them. Cream, as it is obtained from milk, will always contain bacteria in large quantity, and these bacteria will grow as readily in the cream as they will in the milk. The butter maker seldom churns his cream when it is freshly obtained from the milk. There are, it is true, some places where sweet cream butter is made and is in demand, but in the majority of butter-consuming countries a different quality of butter is desired, and the cream is subjected to a process known as "ripening" or "souring" before it is churned. In ripening, the cream is simply allowed to stand in a vat for a period varying from twelve hours to two or three days, according to circumstances. During this period certain changes take place therein. The bacteria which were in the cream originally, get an opportunity to grow, and by the time the ripening is complete they become extremely numerous. As a result, the character of the cream changes just as the milk is changed under similar circumstances. It becomes somewhat soured; it becomes slightly curdled, and acquires a peculiarly pleasant taste and an aroma which was not present in the original fresh cream. After this ripening the cream is churned. It is during the ripening that the bacteria produce their effect, for after the churning they are of less importance. Part of them collect in the butter, part of them are washed off from the butter in the buttermilk and the subsequent processes. Most of the bacteria that are left in the butter soon die, not finding there a favourable condition for growth; some of them, however, live and grow for some time and are prominent agents in the changes by which butter becomes rancid. The butter maker is concerned with the ripening rather than with later processes.

The object of the ripening of cream is to render it in a better condition for butter making. The butter maker has learned by long-experience that ripened cream churns more rapidly than sweet cream, and that he obtains a larger yield of butter therefrom. The great object of the ripening, however, is to develop in the butter the peculiar flavour and aroma which is characteristic of the highest product. Sweet cream butter lacks flavour and aroma, having indeed a taste almost identically the same as cream. Butter, however, that is made from ripened cream has a peculiar delicate flavour and aroma which is well known to lovers of butter, and which is developed during the ripening process.

Bacteriologists have been able to explain with a considerable degree of accuracy the object of this ripening. The process is really a fermentation comparable to the fermentation that takes place in a brewer's malt. The growth of bacteria during the ripening produces chemical changes of a somewhat complicated character, and concerns each of the ingredients of the milk. The lactic-acid organisms affect the milk sugar and produce lactic acid; others act upon the fat, producing slight changes therein; while others act upon the casein and the albumens of the milk. As a result, various biproducts of decomposition arise, and it is these biproducts of decomposition that make the difference between the ripened and the unripened cream. They render it sour and curdle it, and they also produce the flavours and aromas that characterize it. Products of decomposition are generally looked upon as undesirable for food, and this is equally true of these products that arise in cream if the decomposition is allowed to continue long enough. If the ripening, instead of being stopped at the end of a day or two, is allowed to continue several days, the cream becomes decayed and the butter made therefrom is decidedly offensive. But under the conditions of ordinary ripening, when the process is stopped at the right moment, the decomposition products are pleasant rather than unpleasant, and the flavours and aromas which they impart to the cream and to the subsequent butter are those that are desired. It is these decomposition products that give the peculiar character to a high quality of butter, and this peculiar quality is a matter that determines the price which the butter maker can obtain for his product.

But, unfortunately, the butter maker is not always able to depend upon the ripening. While commonly it progresses in a satisfactory manner, sometimes, for no reason that he can assign, the ripening does not progress normally. Instead of developing the pleasant aroma and flavour of the properly ripened cream, the cream develops unpleasant tastes. It may be bitter or somewhat tainted, and just as sure as these flavours develop in the cream, so sure does the quality of the butter suffer. Moreover, it has been learned by experience that some creameries are incapable of obtaining an equally good ripening of their cream. While some of them will obtain favourable results, others, with equal care, will obtain a far less favourable flavour and aroma in their butter. The reason for all this has been explained by modern bacteriology. In the milk, and consequently in the cream, there are always found many bacteria, but these are not always of the same kinds. There are scores, and probably hundreds, of species of bacteria common in and around our barns and dairies, and the bacteria that are abundant and that grow in different lots of cream will not be always the same. It makes a decided difference in the character of the ripening, and in the consequent flavours and aromas, whether one or another species of bacteria has been growing in the cream. Some species are found to produce good results with desired flavours, while others, under identical conditions, produce decidedly poor results with undesired flavours. If the butter maker obtains cream which is filled with a large number of bacteria capable of producing good flavours, then the ripening of his cream will be satisfactory and his butter will be of high quality. If, however, it chances that his cream contains only the species which produce unpleasant flavours, then the character of the ripening will be decidedly inferior and the butter will be of a poorer grade. Fortunately the majority of the kinds of bacteria liable to get into the cream from ordinary sources are such as produce either good effects upon the cream or do not materially influence the flavour or aroma. Hence it is that the ripening of cream will commonly produce good results. Bacteriologists have learned that there are some species of bacteria more or less common around our barns which produce undesirable effects upon flavour, and should these become especially abundant in the cream, then the character of the ripening and the quality of the subsequent butter will suffer. These malign species of bacteria, however, are not very common in properly kept barns and dairies. Hence the process that is so widely used, of simply allowing cream to ripen under the influence of any bacteria that happen to be in it, ordinarily produces good results. But our butter makers sometimes find, at the times when the cattle change from winter to summer or from summer to winter feed, that the ripening is abnormal. The reason appears to be that the cream has become infested with an abundance of malign species. The ripening that they produce is therefore an undesirable one, and the quality of the butter is sure to suffer.

So long as butter was made only in private dairies it was a matter of comparatively little importance if there was an occasional falling off in quality of this sort. When it was made a few pounds at a time, and only once or twice a week, it was not a very serious matter if a few churnings of butter did suffer in quality. But to-day the butter-making industries are becoming more and more concentrated into large creameries, and it is a matter of a good deal more importance to discover some means by which a uniformly high quality can be insured. If a creamery which makes five hundred pounds of butter per day suffers from such an injurious ripening, the quality of its butter will fall off to such an extent as to command a lower price, and the creamery suffers materially. Perhaps the continuation of such a trouble for two or three weeks would make a difference between financial success and failure in the creamery. With our concentration of the butter- making industries it is becoming thus desirable to discover some means of regulating this process more accurately.

The remedy of these occasional ill effects in cream ripening has not been within the reach of the butter maker. The butter maker must make butter with the cream that is furnished him, and if that cream is already impregnated with malign species of bacteria he is helpless. It is true that much can be done to remedy these difficulties by the exercise of especial care in the barns of the patrons of the creamery. If the barns, the cows, the dairies, the milk vessels, etc., are all kept in condition of strict cleanliness, if especial care is taken particularly at the seasons of the year when trouble is likely to arise, and if some attention is paid to the kind of food which the cattle eat, as a rule the cream will not become infected with injurious bacteria. It may be taken as a demonstrated fact that these malign bacteria come from sources of filth, and the careful avoidance of all such sources of filth will in a very large measure prevent their occurrence in the cream. Such measures as these have been found to be practicable in many creameries. Creameries which make the highest priced and the most uniform quality of butter are those in which the greatest care is taken in the barns and dairies to insure cleanliness and in the handling of the milk and cream. With such attention a large portion of the trouble which arises in the creameries from malign bacteria may be avoided.

But these methods furnish no sure remedy against evils of improper species of bacteria in cream ripening, and do not furnish any sure means of obtaining uniform flavour in butter. Even under the very best conditions the flavour of the butter will vary with the season of the year. Butter made in the winter is inferior to that made in the summer months; and while this is doubtless due in part to the different food which the cattle have and to the character of the cream resulting therefrom, these differences in the flavour of the butter are also in part dependent upon the different species of bacteria which are present in the ripening of cream at different seasons. The species of bacteria in June cream are different from those that are commonly present in January cream, and this is certainly a factor in determining the difference between winter and summer butter.

USE OF ARTIFICIAL BACTERIA CULTURES FOR CREAM RIPENING.

Bacteriologists have been for some time endeavouring to aid butter makers in this direction by furnishing them with the bacteria needful for the best results in cream ripening. The method of doing this is extremely simple in principle, but proves to be somewhat difficult in practice. It is only necessary to obtain the species of bacteria that produce the highest results, and then to furnish these in pure culture and in large quantity to the butter makers, to enable them to inoculate their cream with the species of bacteria which will produce the results that they desire. For this purpose bacteriologists have been for several years searching for the proper species of bacteria to produce the best results, and there have been put upon the market for sale several distinct "pure cultures" for this purpose. These have been obtained by different bacteriologists and dairymen in the northern European countries and also in the United States. These pure cultures are furnished to the dairymen in various forms, but they always consist of great quantities of certain kinds of bacteria which experience has found to be advantageous for the purpose of cream ripening.

There have hitherto appeared a number of difficulties in the way of reaching complete success in these directions. The most prominent arises in devising a method of using pure cultures in the creamery. The cream which the butter makers desire to ripen is, as we have seen, already impregnated with bacteria, and would ripen in a fashion of its own even if no pure culture of bacteria were added thereto. Pure cultures can not therefore be used as simply as can yeast in bread dough. It is plain that the simple addition of a pure culture to a mass of cream would not produce the desired effects, because the cream would be ripened then, not by the pure culture alone, but by the pure culture plus all of the bacteria that were originally present. It would, of course, be something of a question as to whether under these conditions the results would be favourable, and it would seem that this method would not furnish any means of getting rid of bad tastes and flavours which have come from the presence of malign species of bacteria. It is plainly desirable to get rid of the cream bacteria before the pure culture is added. This can be readily done by heating it to a temperature of 69 degrees C. (155 degrees F.) for a short time, this temperature being sufficient to destroy most of the bacteria. The subsequent addition of the pure culture of cream-ripening bacteria will cause the cream to ripen under the influence of the added culture alone. This method proves to be successful, and in the butter making countries in Europe it is becoming rapidly adopted.

In this country, however, this process has not as yet become very popular, inasmuch as the heating of the cream is a matter of considerable expense and trouble, and our butter makers have not been very ready to adopt it. For this reason, and also for the purpose of familiarizing butter makers with the use of pure cultures, it has been attempted to produce somewhat similar though less uniform results by the use of pure cultures in cream without previous healing. In the use of pure cultures in this way, the butter maker is directed to add to his cream a large amount of a prepared culture of certain species of bacteria, upon the principle that the addition of such a large number of bacteria to the cream, even though the cream is already inoculated with certain bacteria, will produce a ripening of the cream chiefly influenced by the artificially added culture. The culture thus added, being present in very much greater quantity than the other "wild" species, will have a much greater effect than any of them. This method, of course, cannot insure uniformity. While it may work satisfactorily in many cases, it is very evident that in others, when the cream is already filled with a large number of malign species of bacteria, such an artificial culture would not produce the desired results. This appears to be not only the theoretical but the actual experience. The addition of such pure cultures in many cases produces favourable results, but it does not always do so, and the result is not uniform. While the use of pure cultures in this way is an advantage over the method of simply allowing the cream to ripen normally without such additions, it is a method that is decidedly inferior to that which first pasteurizes the cream and subsequently adds a starter.

There is still another method of adding bacteria to cream to insure a more advantageous ripening, which is frequently used, and, being simpler, is in many cases a decided advantage. This method is by the use of what is called a natural starter. A natural starter consists simply of a lot of cream which has been taken from the most favourable source possible—that is, from the cleanest and best dairy, or from the herd producing the best quality of cream—and allowing this cream to stand in a warm place for a couple of days until it becomes sour. The cream will by that time be filled with large numbers of bacteria, and this is then put as a starter into the vat of cream to be ripened. Of course, in the use of this method the butter maker has no control over the kinds of bacteria that will grow in the starter, but it is found, practically, that if the cream is taken from a good source the results are extremely favourable, and there is produced in this way almost always an improvement in the butter.

The use of pure cultures is still quite new, particularly in this country. In the European butter-making countries they have been used for a longer period and have become very much better known. What the future may develop along this line it is difficult to say; but it seems at least probable that as the difficulties in the details are mastered the time will come when starters will be used by our butter makers for their cream ripening, just as yeast is used by housewives for raising bread, or by brewers for fermenting malt. These starters will probably in time be furnished by bacteriologists. Bacteriology, in other words, is offering in the near future to our butter makers a method of controlling the ripening of the cream in such a way as to insure the obtaining of a high and uniform quality of butter, so far, at least, as concerns flavour and aroma.

BACTERIA IN CHEESE.

Cheese ripening. — The third great product of the dairy industry is cheese, and in connection with this product the dairyman is even more dependent upon bacteria than he is in the production of butter. In the manufacture of cheese the casein of the milk is separated from the other products by the use of rennet, and is collected in large masses and pressed, forming the fresh cheese. This cheese is then set aside for several weeks, and sometimes for months, to undergo a process that is known as ripening. During the ripening there are developed in the cheese the peculiar flavours which are characteristic of the completed product. The taste of freshly made cheese is extremely unlike that of the ripened product. While butter made from unripened cream has a pleasant flavour, and one which is in many places particularly enjoyed, there is nowhere a demand for unripened cheese, for the freshly made cheese has a taste that scarce any one regards as pleasant. Indeed, the whole value of the cheese is dependent upon the flavour of the product, and this flavour is developed during the ripening.

The cheese maker finds in the ripening of his cheese the most difficult part of his manufacture. It is indeed a process over which he has very little control. Even when all conditions seem to be correct, when cheese is made in the most careful manner, it not infrequently occurs that the ripening takes place in a manner that is entirely abnormal, and the resulting cheese becomes worthless. The cheese maker has been at an entire loss to understand these irregularities, nor has he possessed any means of removing them. The abnormal ripening that occurs takes on various types. Sometimes the cheese will become extraordinarily porous, filled with large holes which cause the cheese to swell out of proper shape and become worthless. At other times various spots of red or blue appear in the manufactured cheese; while again unpleasant tastes and flavours develop which render the product of no value. Sometimes a considerable portion of the product of the cheese factory undergoes such irregular ripening, and the product for a long time will thus be worthless. If some means could be discovered of removing these irregularities it would be a great boon to the cheese manufacturer; and very many attempts have been made in one way or another to furnish the cheese maker with some details in the manufacture which will enable him in a measure to control the ripening.

The ripening of the cheese has been subjected to a large amount of study on the part of bacteriologists who have been interested in dairy products. That the ripening of cheese is the result of bacterial growth therein appears to be probable from a priori grounds. Like the ripening of cream, it is a process that occurs somewhat slowly. It is a chemical change which is accompanied by the destruction of proteid matter; it takes place best at certain temperatures, and temperatures which we know are favourable to the growth of micro-organisms, all of which phenomena suggest to us the action of bacteria. Moreover, the flavours and the tastes that arise have a decided resemblance in many cases to the decomposition products of bacteria, strikingly so in Limburger cheese. When we come to study the matter of cheese ripening carefully we learn beyond question that this a priori conclusion is correct. The ripening of any cheese is dependent upon several different factors. The method of preparation, the amount of water left in the curd, the temperature of ripening, and other miscellaneous factors connected with the mechanical process of cheese manufacture, affect its character. But, in addition to all these factors, there is undoubtedly another one, and that is the number and the character of the bacteria that chance to be in the curd when the cheese is made. While it is found that cheeses which are treated by different processes will ripen in a different manner, it is also found that two cheeses which have been made under similar conditions and treated in identically the same way may also ripen in a different manner, so that the resulting flavour will vary. The variations between cheeses thus made may be slight or they may be considerable, but variations certainly do occur. Every one knows the great difference in flavours of different cheeses, and these flavours are due in considerable measure to factors other than the simple mechanical process of making the cheese. The general similarity of the whole process to a bacterial fermentation leads us to believe at the outset that some of the differences in character are due to different kinds of bacteria that multiply in the cheese and produce decomposition therein.

When the matter comes to be studied by bacteriology, the demonstration of this position becomes easy. That the ripening of cheese is due to growth of bacteria is very easily proved by manufacturing cheeses from milk which is deprived of bacteria. For instance, cheeses have been made from milk that has been either sterilized or pasteurized—which processes destroy most of the bacteria therein—and, treated otherwise in a normal manner, are set aside to ripen. These cheeses do NOT ripen, but remain for months with practically the same taste that they had originally. In other experiments the cheese has been treated with a small amount of disinfective, which is sufficient to prevent bacteria from growing, and again ripening is found to be absolutely prevented. Furthermore, if the cheese under ordinary conditions is studied during the ripening process, it is found that bacteria are growing during the whole time. These facts all taken together plainly prove that the ripening of cheese is a fermentation due to bacteria. It will be noticed, however, that the conditions in the cheese are not favourable for very rapid bacterial growth. It is true that there is plenty of food in the cheese for bacterial life, but the cheese is not very moist; it is extremely dense, being subjected in all cases to more or less pressure. The penetration of oxygen into the centre of the mass must be extremely slight. The density, the lack of a great amount of moisture, and the lack of oxygen furnish conditions in which bacteria will not grow very rapidly. The conditions are far less favourable than those of ripening cream, and the bacteria do not grow with anything like the rapidity that they grow in cream. Indeed, the growth of these organisms during the ripening is extremely slow compared to the possibilities of bacterial growth that we have already noticed. Nevertheless, the bacteria do multiply in the cheese, and as the ripening goes on they become more and more abundant, although the number fluctuates, rising and falling under different conditions.

When the attempt is made to determine the relation of the different kinds of ripening to different kinds of bacteria, it has thus far met with extremely little success. That different flavours are due to the ripening produced by different kinds of bacteria would appear to be almost certain when we remember, as we have already noticed, the different kinds of decomposition produced by different species of bacteria. It would seem, moreover, that it ought not to be very difficult to separate from the ripened cheese the bacteria which are present, and thus obtain the kind of bacteria necessary to produce the desired ripening. But for some reason this does not prove to be so easy in practice as it seems to be in theory. Many different species of bacteria have been separated from cheeses. One bacteriologist, studying several cheeses, separated about eighty different species therefrom, and others have found perhaps as many more from different sources. Moreover, experiments have been made with a considerable number of these different kinds of bacteria to determine whether they are capable of producing normal ripening. These experiments consist of making cheese out of milk that has been deprived of its bacteria, and which has been inoculated with large quantities of the species in question. Hitherto these experiments have not been very satisfactory. In some cases the cheese appears to ripen scarcely at all; in other cases the ripening occurs, but the resulting cheese is of a peculiar character, entirely unlike the cheese that it is desired to imitate. There have been one or two experiments in recent times that give a little more promise of success than the earlier ones, for a few species of bacteria have been used in ripening with what the authors have thought to be promising success. The cheese made from the milk artificially inoculated with these species ripens in a satisfactory manner and gives some of the character desired, though up to the present time in no case has the typical normal ripening been produced in any of these experiments.

But these experiments have demonstrated beyond question that the abnormal ripening which is common in cheese factories is due to the presence of undesirable species of bacteria in the milk. Many of the experiments in making cheeses by means of artificial cultures of bacteria have resulted in decidedly abnormal cheeses. Many of the cheeses thus manufactured have shown imperfections in ripening which are identical with those actually occurring in the cheese factory. Several different species of bacteria have been found which, when artificially used thus for ripening cheese, will give rise to the porosity and the abnormal swelling of the cheese already referred to (Fig. 24). Others produced bad tastes and flavours, and enough has been done in this line to demonstrate beyond peradventure that the abnormal ripening of cheese is due primarily to the growth of improper species therein. Quite a long list of species of bacteria which produce abnormal ripening have been isolated from cheeses, and have been studied and experimented with by bacteriologists. As a result of this study of abnormal ripening, there has been suggested a method of partially controlling these—remedying them. The method consists simply in testing the fermenting qualities of the milk used. A small sample of milk from different dairies is allowed to stand in the cheese factory by itself until it undergoes its normal souring. If the fermentation or souring that thus occurs is of a normal character, the milk is regarded as proper for cheese making. But if the fermentation that occurs in any particular sample of milk is unusual; if an extraordinary amount of gas bubbles are produced, or if unpleasant smells and tastes arise, the sample is regarded as unfavourable for cheese making, and as likely to produce abnormal ripening in the cheeses. Milk from this source would therefore be excluded from the milk that is to be used in cheese making. This, of course, is a tentative and an unsatisfactory method of controlling the ripening, and yet it is one of some practical value to cheese makers. It is the only method that has yet been suggested of controlling the ripening.

Our bacteriologists, of course, are quite confident that in the future more practical results will be obtained along this line than in the past. If it is true that cheeses are ripened by bacteria; if it is true that different qualities in the cheese are due to the growth of different species of bacteria during the ripening, it would seem to be possible to obtain the proper kind of bacteria and to furnish them to the cheese maker for artificially inoculating his cheese, just as it has been possible to furnish artificially cultivated yeasts to the brewer, and as it has become possible to furnish artificially cultivated bacteria to the butter maker. We must, however, recognise this to be a matter for the future. Up to the present time no practical results along the lines of bacteria have been obtained which our cheese manufacturers can make use of in the way of controlling with any accuracy this process of cheese ripening.

Thus it will be seen that in this last dairy product bacteria play even a more important part than in any of the others. The food value of cheese is dependent upon the casein which is present. The market price, however, is controlled entirely by the flavour, and this flavour is a product of bacterial growth. Upon the action of bacteria, then, the cheese maker is absolutely dependent; and when our bacteriologists are able in the future to investigate this matter further, it seems to be at least possible that they may obtain some means of enabling the cheese maker to control the ripening accurately. Not only so, but recognising the great variety in the flavours of cheese, and recognising that different kinds of bacteria undoubtedly produce different kinds of decomposition products, it seems to be at least possible that a time will come when the cheese maker will be able to produce at— will any particularly desired flavour in his cheese by the addition to it of particular species of bacteria, or particular mixtures of species of bacteria which have been discovered to produce the desired effects.

CHAPTER IV.

BACTERIA IN NATURAL PROCESSES. — AGRICULTURE.

Thus far, in considering the relations of bacteria to mankind, we have taken into account only the arts and manufactures, and have found bacteria playing no unimportant part in many of the industries of our modern civilized life. So important are they that there is no one who is not directly affected by them. There is hardly a moment in our life when we are not using some of the direct or indirect products of bacterial action. We turn now, however, to the consideration of a matter of even more fundamental importance; for when we come to study bacteria in Nature, we find that there are certain natural processes connected with the life of animals and plants that are fundamentally based upon their powers. Living Nature appears limitless, for life processes have been going on in the world through countless centuries with seemingly unimpaired vigour. At the very bottom we find this never-ending exhibition of vital power dependent upon certain activities of micro-organisms. So thoroughly is this true that, as we shall find after a short consideration, the continuance of life upon the surface of the world would be impossible if bacterial action were checked for any considerable length of time. The life of the globe is, in short, dependent upon these micro-organisms.

BACTERIA AS SCAVENGERS.

In the first place, we may notice the value of these organisms simply as scavengers, keeping the surface of the earth in the proper condition for the growth of animals and plants. A large tree in the forest dies and falls to the ground. For a while the tree trunk lies there a massive structure, but in the course of months a slow change takes place in it. The bark becomes softened and falls from the wood. The wood also becomes more or less softened; it is preyed and more members are required to complete the chain. The transference of matter through a series of changes by which it is brought from a condition in which it is proper food for plants back again into a condition when it is once more a proper food for plants, is one of the interesting discoveries of modern science, and one in which, as we shall see, bacteria play a most important part. This food cycle is illustrated roughly by the accompanying diagram; but in order to understand it, an explanation of the various steps in this cycle is necessary.

It will be noticed that at the bottom of the circle represented in Fig. 25, at A, are given various ingredients which are found in the soil and which form plant foods. Plant foods, as may be seen there, are obtained partly from the air as carbonic dioxide and water; but another portion comes from the soil. Among the soil ingredients the most prominent are nitrates, which are the forms of nitrogen compounds most easily made use of by plants as a source of this important element. It should be stated also that there are other compounds in the soil which furnish plants with part of their food — compounds containing potassium, phosphorus, and some other elements. For simplicity's sake, however, these will be left out of consideration. Beginning at the bottom of the cycle (Fig. 25 A), plant life seizes the gases from the air and these foods from the soil, and by means of the energy furnished it by the sun's rays builds these simple chemical compounds into more complex ones. This gives us the second step, as shown in Fig. 25 B, the products of plant life. These products of plant life consist of such materials as sugar, starches, fats, and proteids, all of which have been manufactured by the plant from the ingredients furnished it from the soil and air, and through the agency of the sun's rays. These products of plant life now form foods for the animal kingdom. Starches, fats, and proteids are animal foods, and upon such complex bodies alone can the animal kingdom be fed. Animal life, standing high up in the circle, is not capable of extracting its nutriment from the soil, but must take the more complex foods which have been manufactured by plant life. These complex foods enter now into the animal and take their place in the animal body. By the animal activities, some of the foods are at once decomposed into carbonic acid and water, which, being dissipated into the air, are brought back at once into the condition in which they can serve again as plant food. This part of the food is thus brought back again to the bottom of the circle (Fig. 25, dotted lines). But while it is true that animals do thus reduce some of their foods to the simple condition of carbonic acid and water, this is not true of most of the foods which contain nitrogen. The nitrogenous foods are as necessary for the life as the carbon foods, and animals do not reduce their nitrogenous foods to the condition in which plants can prey upon them. While plants furnish them with nitrogenous food, they can not give it back to the plants. Part of the nitrogenous foods animals build into new albumins (Fig. 25 C); but a part of them they reduce at once into a somewhat simpler condition known as urea. Urea is the form in which the nitrogen is commonly excreted from the animal body. But urea is not a plant food; for ordinary plants are entirely unable to make use of it. Part of the nitrogen eaten by the animal is stored up in its body, and thus the body of the animal, after it has died, contains these nitrogen compounds of high complexity. But plants are not able to use these compounds. A plant can not be fed upon muscle tissue, nor upon fats, nor bones, for these are compounds so complex that the simple plant is unable to use them at all. So far, then, in the food cycle the compounds taken from the soil have been built up into compounds of greater and greater complexity; they have reached the top of this circle, and no part of them, except part of the carbon and oxygen, has become reduced again to plant food. In order that this material should again become capable of entering into the life of plants so as to go over the circle again, it is necessary for it to be once more reduced from its highly complex condition into a simpler one.

Now come into play these decomposition agencies which we have been studying under the head of scavengers. It will be noticed that the next step in the food cycle is taken by the decomposition bacteria. These organisms, existing, as we have already seen, in the air, in the soil, in the water, and always ready to seize hold of any organic substance that may furnish them with food, feed upon the products of animal life, whether they are such products as muscle tissue, or fat, or sugar, or whether they are the excreted products of animal life, such as urea, and produce therein the chemical decomposition changes already noticed. As a result of this chemical decomposition, the complex bodies are broken into simpler and simpler compounds, and the final result is a very thorough destruction of the animal body or the material excreted by animal life, and its reduction into forms simple enough for plants to use again as foods. Thus the bacteria come in as a necessary link to connect the animal body, or the excretion from the animal body, with the soil again, and therefore with that part of the circle in which the material can once more serve as plant food.

But in the decomposition that thus occurs through the agency of the putrefactive bacteria it very commonly happens that some of the food material is broken down into compounds too simple for use as plant food. As will be seen by a glance at the diagram (Fig. 25 D), a portion of the cleavage products resulting from the destruction of these animal foods takes the form of carbonic-acid gas and water. These ingredients are at once in condition for plant life, as shown by the dotted lines. They pass off into the air, and the green leaves of vegetation everywhere again seize them, assimilate them, and use them as food. Thus it is that the carbon and the oxygen have completed the cycle, and have come back again to the position in the circle where they started. In regard to the nitrogen portion of the food, however, it very commonly happens that the products which arise as the result of the decomposition processes are not yet in proper condition for plant food. They are reduced into a condition actually too simple for the use of plants. As a result of these putrefactive changes, the nitrogen products of animal life are broken frequently into compounds as simple as ammonia (NH_3), or into compounds which the chemists speak of as nitrites (Fig. 25 at D). Now these compounds are not ordinarily within the reach of plant life. The luxuriant vegetation of the globe extracts its nitrogen from the soil in a form more complex than either of the compounds here mentioned; for, as we have seen, it is nitrates chiefly that furnish plants with their nitrogen food factor. But nitrates contain considerable oxygen. Ammonia, which is one of the products of putrefactive de- composition, contains no oxygen, and nitrites, another factor, contains less oxygen than nitrates. These bodies are thus too simple for plants to make use of as a source of nitrogen. The chemical destruction of the food material which results from the action of the putrefactive bacteria is too thorough, and the nitrogen foods are not yet in condition to be used by plants.

Now comes in the agency of still another class of microorganisms, the existence of which has been demonstrated to us during the last few years. In the soil everywhere, especially in fertile soil, is a class of bacteria which has received the name of nitrifying bacteria (Fig. 26). These organisms grow in the soil and feed upon the soil ingredients. In the course of their life they have somewhat the same action upon the simple nitrogen cleavage products just mentioned as we have already noticed the vinegar- producing species have upon alcohol, viz., the bringing about a union with oxygen. There are apparently several different kinds of nitrifying bacteria with different powers. Some of them cause an oxidation of the nitrogen products by means of which the ammonia is united with oxygen and built up into a series of products finally resulting in nitrates (Fig. 26). By the action of other species still higher nitrogen compounds, including the nitrites, are further oxidized and built up into the form of nitrates. Thus these nitrifying organisms form the last link in the chain that binds the animal kingdom to the vegetable kingdom (Fig. 25 at 4). For after the nitrifying organisms have oxidized nitrogen cleavage products, the results of the oxidation in the form of nitrates or nitric acid are left in the soil, and may now be seized upon by the roots of plants, and begin once more their journey around the food cycle. In this way it will be seen that while plants, by building up compounds, form the connecting link between the soil and animal life, bacteria in the other half of the cycle, by reducing them again, give us the connecting link between animal life and the soil. The food cycle would be as incomplete without the agency of bacterial life as it would be without the agency of plant life.

But even yet the food cycle is not complete. Some of the processes of decomposition appear to cause a portion of the nitrogen to fly out of the circle at a tangent. In the process of decomposition which is going on through the agency of micro-organisms, a considerable part of the nitrogen is dissipated into the air in the form of free nitrogen. When a bit of meat decays, part of the meat is, indeed, converted into ammonia or other nitrogen compounds, but if the putrefaction is allowed to go on, in the end a considerable portion of it will be broken into still simpler forms, and the nitrogen will finally be dissipated into the air in the form of free nitrogen. This dissipation of free nitrogen into the air is going on in the world wherever putrefaction takes place. Wherever decomposition of nitrogen products occurs some free nitrogen is eliminated. Now, this part of the nitrogen has passed beyond the reach of plants, for plants can not extract free nitrogen from the air. In the diagram this is represented as a portion of the material which, through the agency of the decomposition bacteria, has been thrown out of the cycle at a tangent (Fig. 25 E). It will, of course, be plain from this that the store of nitrogen food must be constantly diminishing. The soil may have been originally supplied with a given quantity of nitrogen compound, but if the decomposition products are causing considerable quantities of this nitrogen to be dissipated in the air, it plainly follows that the total amount of nitrogen food upon which the animal and vegetable kingdoms can depend is becoming constantly reduced by such dissipation.

There are still other methods by which nitrogen is being lost from the food cycle. First, we may notice that the ordinary processes of vegetation result in a gradual draining of the soil and a throwing of its nitrogen into the ocean. The body of any animal or any plant that chances to fall into a brook or river is eventually carried to the sea, and the products of its decomposition pass into the ocean and are, of course, lost to the soil. Now, while this gradual extraction of nitrogen from the soil by drainage is a slow one, it is nevertheless a sure one. It is far more rapid in these years of civilized life than in former times, since the products of the soil are given to the city, and then are thrown into its sewage Our cities, then, with our present system of disposing of sewage, are draining from the soil the nitrogen compounds and throwing them away.

In yet another direction must it be noticed that our nitrogen compounds are being lost to plant life—viz., by the use of various nitrogen compounds to form explosives. Gunpowder, nitro-glycerine, dynamite, in fact, nearly all the explosives that are used the world over for all sorts of purposes, are nitrogen compounds. When they are exploded the nitrogen of the compound is dissipated into the air in the form of gas, much of it in the form of free nitrogen. The basis from which explosive compounds are made contains nitrogen in the form in which it can be used by plants. Saltpetre, for example, is equally good as a fertilizer and as a basis for gunpowder. The products of the explosion are gases no longer capable of use by plants, and thus every explosion of nitrogen compounds aids in this gradual dissipation of nitrogen products, taking them from the store of plant foods and throwing them away.

All of these agencies contribute to reduce the amount of material circulating in the food cycle of Nature, and thus seem to tend inevitably in the end toward a termination of the processes of life; for as soon as the soil becomes exhausted of its nitrogen compounds, so soon will plant life cease from lack of nutrition, and the disappearance of animal life will follow rapidly. It is this loss of nitrogen in large measure that is forcing our agriculturists to purchase fertilizers. The last fifteen years have shown us, however, that here again we may look upon our friends, the bacteria, as agents for counteracting this dissipating tendency in the general processes of Nature. Bacterial life in at least two different ways appears to have the function of reclaiming from the atmosphere more or less of this dissipated free nitrogen.

In the first place, it has been found in the last few years that soil entirely free from all common plants, but containing certain kinds of bacteria, if allowed to stand in contact with the air, will slowly but surely gain in the amount of nitrogen compounds that it contains. These nitrogen compounds are plainly manufactured by the bacteria in the soil; for unless the bacteria are present they do not accumulate, and they do accumulate inevitably if the bacteria are present in the proper quantity and the proper species. It appears that, as a rule, this fixation of nitrogen is not performed by any one species of microorganisms, but by two or three of them acting together. Certain combinations of bacteria have been found which, when inoculated in the soil, will bring about this fixation of nitrogen, but no one of the species is capable of producing this result alone. We do not know to what extent these organisms are distributed in the soil, nor how widely this nitrogen fixation through bacterial life is going on. It is only within a short time that it has been demonstrated to exist, but we must look upon bacteria in the soil as one of the factors in reclaiming from the atmosphere the dissipated free nitrogen.

The second method by which bacteria aid in the reclaiming of this lost nitrogen is by a combined action of certain species of bacteria and some of the higher plants. Ordinary green plants, as already noted, are unable to make use of the free nitrogen of the atmosphere It was found, however, some fifteen years ago that some species of plants, chiefly the great family of legumes, which contains the pea plant, the bean, the clover, etc, are able, when growing in soil that is poor in nitrogen, to obtain nitrogen from some source other than the soil in which they grow. A pea plant in soil that contains no nitrogen products and watered with water that contains no nitrogen, will, after sprouting and growing for a length of time, be found to have accumulated a considerable quantity of fixed nitrogen in its tissues The only source of this nitrogen has been evidently from the air which bathes the leaves of the plant or permeates the soil and bathes its roots This fact was at first disputed, but subsequently demonstrated to be true, and was found later to be associated with the combined action of these legumes and certain soil bacteria. When a legume thus gains nitrogen from the air, it develops upon its roots little bunches known as root nodules or root tubercles. The nodules are sometimes the size of the head of a pm, and sometimes much larger than this, occasionally reaching the size of a large pea, or even larger. Upon microscopic examination they are found to be little nests of bacteria In some way the soil organisms (Fig 27) make their way into the roots of the sprouting plant, and finding there congenial environment, develop in considerable quantities and produce root tubercles in the root. Now, by some entirely unknown process, the legume and the bacteria growing together succeed in extracting the nitrogen from the atmosphere which permeates the soil, and fixing this nitrogen in the tubercles and the roots in the form of nitrogen compounds. The result is that, after a proper period of growth, the amount of fixed nitrogen in the plant is found to have very decidedly increased (Fig 25 E).

This, of course, furnishes a starting point for the reclaiming of the lost atmospheric nitrogen. The legume continues to live its usual life, perhaps increasing the store of nitrogen in its roots and stems and leaves during the whole of its normal growth. Subsequently, after having finished its ordinary life, the plant will die, and then the roots and stems and leaves, falling upon the ground and becoming buried, will be seized upon by the decomposition bacteria already mentioned. The nitrogen which has thus become fixed in their tissues will undergo the destructive changes already described. This will result eventually in the production of nitrates. Thus some of the lost nitrogen is restored again to the soil in the form of nitrates, and may now start on its route once more around the cycle of food.

It will be seen, then, that the food cycle is a complete one. Beginning with the mineral ingredients in the soil, the food matter may start on its circulation from the soil to the plant, from the plant to the animal, from the animal to the bacterium and from the bacterium through a series of other bacteria back again to the soil in the condition in which it started. If, perchance, in this progress around the circle some of the nitrogen is thrown off at a tangent, this, too, is brought back again to the circle through the agency of bacterial life. And so the food material of animals and plants continues in this never-ceasing circulation. It is the sunlight that furnishes the energy for the motion. It is the sunlight that forces the food around the circle and keeps up the endless change; and so long as, the sun continues to shine upon the earth there seems to be no reason why the process should ever cease. It is this repeated circulation that has made the continuation of life possible for the millions and millions of years of the earth's history. It is this continued circulation that makes life possible still, and it is only this fact that the food is thus capable of ever circulating from animal to plant and from plant to animal that makes it possible for the living world to continue its existence. But, ah we have seen, one half of this great circle of food change is dependent upon bacterial life. Without the bacterial life the animal body and the animal excretion could never be brought back again within the reach of the plant; and thus, were it not for the action of these micro- organisms the food cycle would be incomplete and life could not continue indefinitely upon the surface of the earth. At the very foundation, the continuation of the present condition of Nature and the existence of life during the past history of the world has been fundamentally based upon the ubiquitous presence of bacteria and upon their continual action in connection with both destructive and constructive processes.

RELATION OF BACTERIA TO AGRICULTURE.

We have already noticed that bacteria play an important part in some of the agricultural industries, particularly in the dairy. From the consideration of the matters just discussed, it is manifest that these organisms must have an even more intimate relation to the farmer's occupation. At the foundation, farming consists in the cultivation of plants and animals, and we have already seen how essential are the bacteria in the continuance of animal and plant life. But aside from these theoretical considerations, a little study shows that in a very practical manner the farmer is ever making use of bacteria, as a rule, quite unconsciously, but none the less positively.

SPROUTING OF SEEDS.

Even in the sprouting of seeds after they are sown in the soil bacterial life has its influence. When seeds are placed m moist soil they germinate under the influence of heat. The rich albuminous material in the seeds furnishes excellent food, and inasmuch as bacteria abound in the soil, it is inevitable that they should grow in and feed upon the seed. If the moisture is excessive and the heat considerable, they very frequently grow so rapidly in the seed as to destroy its life as a seedling. The seed rots in the ground as a result. This does not commonly occur, however, in ordinary soil. But even here bacteria do grow in the seed, though not so abundantly as to produce any injury. Indeed, it has been claimed that their presence in the seed in small quantities is a necessity for the proper sprouting of the seed. It has been claimed that their growth tends to soften the food material in the seed, so that the young seedling can more readily absorb it for its own food, and that without such a softening the seed remains too hard for the plant to use. This may well be doubted, however, for seeds can apparently sprout well enough without the aid of bacteria. But, nevertheless, bacteria do grow in the seed during its germination, and thus do aid the plant in the softening of the food material. We can not regard them as essential to seed germination. It may well be claimed that they ordinarily play at least an incidental part in this fundamental life process, although it is uncertain whether the growth of seedlings is to any considerable extent aided thereby.

THE SILO.

In the management of a silo the farmer has undoubtedly another great bacteriological problem. In the attempt to preserve his summer-grown food for the winter use of his animals, he is hindered by the activity of common bacteria. If the food is kept moist, it is sure to undergo decomposition and be ruined in a short time as animal food. The farmer finds it necessary, therefore, to dry some kinds of foods, like hay. While he can thus preserve some foods, others can not be so treated. Much of the rank growth of the farm, like cornstalks, is good food while it is fresh, but is of little value when dried. The farmer has from experience and observation discovered a method of managing bacterial growth which enables him to avoid their ordinary evil effects. This is by the use of the silo. The silo is a large, heavily built box, which is open only at the top. In the silo the green food is packed tightly, and when full all access of air is excluded, except at its surface. Under these conditions the food remains moist, but nevertheless does not undergo its ordinary fermentations and putrefactions, and may be preserved for months without being ruined. The food in such a silo may be taken out months after it is packed, and will still be found to be in good condition for food. It is true that it has changed its character somewhat, but it is not decayed, and is eagerly eaten by cattle.

We are yet very ignorant of the nature of the changes which occur m the food while in the silo. The food is not preserved from fermentation. When the siloxis packed slowly, a very decided fermentation occurs by which the mass is raised to a high temperature (140 degrees F. to 160 degrees F.). This heating is produced by certain species of bacteria which grow readily even at this high temperature. The fermentation uses up the air in the silo to a certain extent and produces a settling of the material which still further excludes air. The first fermentation soon ceases, and afterward only slow changes occur. Certain acid- producing bacteria after a little begin to grow slowly, and in time the silage is rendered somewhat sour by the production of acetic acid. But the exclusion of air, the close packing, and the small amount of moisture appear to prevent the growth of the common putrefactive bacteria, and the silage remains good for a long time. In other methods of filling the silo, the food is very quickly packed and densely crowded together so as to exclude as much air as possible from the beginning. Under these conditions the lack of moisture and air prevents fermentative action very largely. Only certain acid-producing organisms grow, and these very slowly. The essential result in either case is that the common putrefactive bacteria are prevented from growing, probably by lack of sufficient oxygen and moisture, and thus the decay is prevented. The closely packed food offers just the same unfavourable condition for the growth of common putrefactive bacteria that we have already seen offered by the hard-pressed cheese, and the bacteria growth is in the same way held in check. Our knowledge of the matter is as yet very slight, but we do know enough to understand that the successful management of a silo is dependent upon the manipulation of bacteria.

THE FERTILITY OF THE SOIL.

The farmer's sole duty is to extract food from the soil. This he does either directly by raising crops, or indirectly by raising animals which feed upon the products of the soil. In either case the fertility of the soil is the fundamental factor in his success. This fertility is a gift to him from the bacteria.

Even in the first formation of soil he is in a measure dependent upon bacteria. Soil, as is well known, is produced in large part by the crumbling of the rocks into powder. This crumbling we generally call weathering, and regard it as due to the effect of moisture and cold upon the rocks, together with the oxidizing action of the air. Doubtless this is true, and the weathering action is largely a physical and chemical one. Nevertheless, in this fundamental process of rock disintegration bacterial action plays a part, though perhaps a small one. Some species of bacteria, as we have seen, can live upon very simple foods, finding in free nitrogen and carbonates sufficiently highly complex material for their life. These organisms appear to grow on the bare surface of rocks, assimilating nitrogen from the air, and carbon from some widely diffused carbonates or from the CO_2 in the air. Their secreted products of an acid nature help to soften the rocks, and thus aid in performing the first step in weathering.

The soil is not, however, all made up of disintegrated rocks. It contains, besides, various ingredients which combine to make it fertile. Among these are various sulphates which form important parts of plant foods. These sulphates appear to be formed, in part, at least, by bacterial agency. The decomposition of proteids gives rise, among other things, to hydrogen sulphide (H2S). This gas, which is of common occurrence in the atmosphere, is oxidized by bacterial growth into sulphuric acid, and this is the basis of part of the soil sulphates. The deposition of iron phosphates and iron silicates is probably also in a measure aided by bacterial action. All of these processes are factors in the formation of soil. Beyond much question the rock disintegration which occurs everywhere in Nature is chiefly the result of physical and chemical changes, but there is reason for believing that the physical and chemical processes are, to a slight extent at least, assisted by bacterial life.

A more important factor of soil fertility is its nitrogen content, without which it is completely barren. The origin of these nitrogen ingredients has been more or less of a puzzle. Fertile soil everywhere contains nitrates and other nitrogen compounds, and in certain parts of the world there are large accumulations of these compounds, like the nitrate beds of Chili. That they have come ultimately from the free atmospheric nitrogen seems certain, and various attempts have been made to explain a method of this nitrogen fixation. It has been suggested that electrical discharges in the air may form nitric acid, which would readily then unite with soil ingredients to form nitrates. There is little reason, however, for believing this to be a very important factor But in the soil bacteria we find undoubtedly an efficient agency m this nitrogen fixation. As already seen, the bacteria are able to seize the free atmospheric nitrogen, converting it into nitrite and nitrates. We have also learned that they can act in connection with legumes and some other plants, enabling them to fix atmospheric nitrogen and store it m their roots. By these two means the nitrogen ingredient in the soil is prevented from becoming exhausted by the processes of dissipation constantly going on. Further, by some such agency must we imagine the original nitrogen soil ingredient to have been derived. Such an organic agency is the only one yet discerned which appears to have been efficient in furnishing virgin soil with its nitrates, and we must therefore look upon bacteria as essential to the original fertility of the soil. But in another direction still does the farmer depend directly upon bacteria The most important factor in the fertility of the soil is the part of it called humus. This humus is very complex, and never alike in different soils It contains nitrogen compounds in abundance, together with sulphates, phosphates, sugar, and many other substances. It is this which makes the garden soil different from sand, or the rich soil different from the sterile soil. If the soil is cultivated year after year, its food ingredients are slowly but surely exhausted. Something is taken from the humus each year, and unless this be replaced the soil ceases to be able to support life. To keep up a constant yield from the soil the farmer understands that he must apply fertilizers more or less constantly.

This application of fertilizers is simply feeding the crops. Some of these fertilizers the farmer purchases, and knows little or nothing as to their origin. The most common method of feeding the crops is, however, by the use of ordinary barnyard manure. The reason why this material contains plant food we can understand, since it is made of the undigested part of food, together with all the urea and other excretions of animals, and contains, therefore, besides various minerals, all of the nitrogenous waste of animal life. These secretions are not at first fit for plant food. The farmer has learned by experience that such excretions, before they are of any use on his fields, must undergo a process of slow change, which is sometimes called ripening. Fresh manure is sometimes used on the fields, but it is only made use of by the plants after the ripening process has occurred. Fresh animal excretions are of little or no value as a fertilizer. The farmer, therefore, commonly allows it to remain in heaps for some time, and it undergoes a slow change, which gradually converts it into a condition in which it can be used by plants. This ripening is readily explained by the facts already considered The fresh animal secretions consist of various highly complex compounds of nitrogen, and the ripening is a process of their decomposition. The proteids are broken to pieces, and their nitrogen elements reduced to the form of nitrates, leucin, etc, or even to ammonia or free nitrogen. Further, a second process occurs, the process of oxidation of these nitrogen compounds already noticed, and the ammonia and nitrites resulting from the decomposition are built into nitrates. In short, in this ripening manure the processes noticed in the first part of this chapter are taking place, by which the complex nitrogenous bodies are first reduced and then oxidized to form plant food. The ripening of manure is both an analytical and a synthetical process. By the analysis, proteids and other bodies are broken into very simple compounds, some of them, indeed, being dissipated into the air, but other portions are retained and then oxidized, and these latter become the real fertilizing materials. Through the agency of bacteria the compost heap thus becomes the great source of plant food to the farmer. Into this compost heap he throws garbage, straw, vegetable and animal substances in general, or any organic refuse which may be at hand. The various bacteria seize it all, and cause the decomposition which converts it into plant food again. The rotting of the compost heap is thus a gigantic cultivation of bacteria.

This knowledge of the ripening process is further teaching the farmer how to prevent waste. In the ordinary decomposition of the compost heap not an inconsiderable portion of the nitrogen is lost in the air by dissipation as ammonia or free nitrogen. Even his nitrates may be thus lost by bacterial action. This portion is lost to the farmer completely, and he can only hope to replace it either by purchasing nitrates in the form of commercial fertilizers, or by reclaiming it from the air by the use of the bacterial agencies already noticed. With the knowledge now at his command he is learning to prevent this waste. In the decomposition one large factor of loss is the ammonia, which, being a gas, is readily dissipated into the air. Knowing this common result of bacterial action, the scientist has told the farmer that, by adding certain common chemicals to his decomposing manure heap, chemicals which will readily unite with ammonia, he may retain most of the nitrogen in this heap in the form of ammonia salts, which, once formed, no longer show a tendency to dissipate into the air. Ordinary gypsum, or superphosphates, or plaster will readily unite with ammonia, and these added to the manure heap largely counteract the tendency of the nitrogen to waste, thus enabling the farmer to put back into his soil most of the nitrogen which was extracted from it by his crops and then used by his stock. His vegetable crops raise the nitrates into proteids. His animals feed upon the proteids, and perform his work or furnish him with milk. Then his bacteria stock take the excreted or refuse nitrogen, and in his manure heap turn it back again into nitrates ready to begin the circle once more. This might go on almost indefinitely were it not for two facts, the farmer sends nitrogenous material off his farm in the milk or grains or other nitrogenous products, which he sells, and the decomposition processes, as we have seen, dissipate some of the nitrogen into the air as free nitrogen.

To meet this emergency and loss the farmer has another method of enriching the soil, again depending upon bacteria. This is the so-called green manuring. Here certain plants which seize nitrogen from the air are cultivated upon the field to be fertilized, and, instead of harvesting a crop, it is ploughed into the soil. Or perhaps the tops may be harvested, the rest being ploughed into the soil. The vegetable material thus ploughed in lies over a season and enriches the soil. Here the bacteria of the soil come into play in several directions. First, if the crop sowed be a legume, the soil bacteria assist it to seize the nitrogen from the air. The only plants which are of use in this green manuring are those which can, through the agency of bacteria, obtain nitrogen from the air and store it in their roots. Second, after the crop is ploughed into the soil various decomposing bacteria seize upon it, pulling the compounds to pieces. The carbon is largely dissipated into the air as carbonic dioxide, where the next generation of plants can get hold of it. The minerals and the nitrogen remain in the soil. The nitrogenous portions go through the same series of decomposition and synthetical changes already described, and thus eventually the nitrogen seized from the air by the combined action of the legumes and the bacteria is converted into nitrates, and will serve for food for the next set of plants grown on the same soil. Here is thus a practical method of using the nitrogen assimilation powers of bacteria, and reclaiming nitrogen from the air to replace that which has been lost. Thus it is that the farmer's nitrogen problem of the fertile soil appears to resolve itself into a proper handling of bacteria. These organisms have stocked his soil in the first place. They convert all of his compost heap wastes into simple bodies, some of which are changed into plant foods, while others are at the same time lost. Lastly, they may be made to reclaim this lost nitrogen, and the fanner, so soon as he has requisite knowledge of these facts, will be able to keep within his control the supply of this important element. The continued fertility of the soil is thus a gift from the bacteria.

BACTERIA AS SOURCES OF TROUBLE TO THE FARMER.

While the topics already considered comprise the most important factors in agricultural bacteriology, the farmer's relations to bacteria do not end here. These organisms come incidentally into his life in many ways. They are not always his aids as they are in most of the instances thus far cited. They produce disease in his cattle, as will be noticed in the next chapter. Bacteria are agents of decomposition, and they are just as likely to decompose material which the farmer wishes to preserve as they are to decompose material which the farmer desires to undergo the process of decay. They are as ready to attack his fruits and vegetables as to ripen his cream. The skin of fruits and vegetables is a moderately good protection of the interior from the attack of bacteria; but if the skin be broken in any place, bacteria get in and cause decay, and to prevent it the farmer uses a cold cellar. The bacteria prevent the farmer from preserving meats for any length of time unless he checks their growth in some way. They get into the eggs of his fowls and ruin them. Their troublesome nature in the dairy in preventing the keeping of milk has already been noticed. If he plants his seeds in very moist, damp weather, the soil bacteria cause too rapid a decomposition of the seeds and they rot in the ground instead of sprouting. They produce disagreeable odours, and are the cause of most of the peculiar smells, good and bad, around the barn. They attack the organic matter which gets into his well or brook or pond, decomposing it, filling the water with disagreeable and perhaps poisonous products which render it unfit to drink. They not only aid in the decay of the fallen tree in his forests; but in the same way attack the timber which he wishes to preserve, especially if it is kept in a moist condition. Thus they contribute largely to the gradual destruction of wooden structures. It is therefore the presence of these organisms which forces him to dry his hay, to smoke his hams, to corn his beef, to keep his fruits and vegetables cool and prevent skin bruises, to ice his dairy, to protect his timber from rain, to use stone instead of wooden foundations for buildings, etc. In general, when the farmer desires to get rid of any organic refuse, he depends upon bacteria, for they are his sole agents (aside from fire) for the final destruction of organic matter. When he wishes to convert waste organic refuse into fertilizing material, he uses the bacteria of his compost heap. On the other hand, whenever he desires to preserve organic material, the bacteria are the enemies against which he must carefully guard.

Thus the farmer's life from year's end to year's end is in most intimate association with bacteria. Upon them he depends to insure the continued fertility of his soil and the constant continued production of good crops. Upon them he depends to turn into plant food all the organic refuse from his house or from his barn. Upon them he depends to replenish his stock of nitrogen. It is these organisms which furnish his dairy with its butter flavours and with the taste of its cheese. But, on the other hand, against them he must be constantly alert. All his food products must be protected from their ravages. A successful farmer's life, then, largely resolves itself into a skilful management of bacterial activity. To aid them in destroying or decomposing everything which he does not desire to preserve, and to prevent their destroying the organic material which he wishes to keep for future use, is the object of a considerable portion of farm labour; and the most successful farmer to-day, and we believe the most successful farmer of the future, is the one who most intelligently and skilfully manipulates these gigantic forces furnished him by the growth of his microscopical allies.

RELATION OF BACTERIA TO COAL. Another one of Nature's processes in which bacteria have played an important part is in the formation of coal. It is unnecessary to emphasize the importance of coal in modern civilization. Aside from its use as fuel, upon which civilization is dependent, coal is a source of an endless variety of valuable products. It is the source of our illuminating gas, and ammonia is one of the products of the gas manufacture. From the coal also comes coal tar, the material from which such a long series of valuable materials, as aniline colours, carbolic acid, etc, is derived. The list of products which we owe to coal is very long, and the value of this material is hardly to be overrated. In the preparation of these ingredients from coal bacteria do not play any part. Most of them are derived by means of distillation. But when asked for the agents which have given us the coal of the coal beds, we shall find that here, too, we owe a great debt to bacteria.

Coal, as is well known, has come from the accumulation of the luxuriant vegetable growth of the past geological ages. It has therefore been directly furnished us by the vegetation of the green plants of the past, and, in general, it represents so much carbonic dioxide which these plants have extracted from the atmosphere. But while the green plants have been the active agents in producing this assimilation, bacteria have played an important part in coal manufacture in two different directions. The first appears to be in furnishing these plants with nitrogen. Without a store of fixed nitrogen in the soil these carboniferous plants could not have grown. This matter has already been considered. We have no very absolute knowledge as to the agency of bacteria in furnishing nitrogen for this vegetation in past ages, but there is every reason to believe that in the past, as in the present, the chief source of organic nitrogen has been from the atmosphere and derived from the atmosphere through the agency of bacteria. In the absence of any other known factor we may be pretty safe in the assumption that bacteria played an important part in this nitrogen fixation, and that bacteria must therefore be regarded as the agents which have furnished us the nitrogen stored in the coal.

But in a later stage of coal formation bacteria have contributed more directly to the formation of coal. Coal is not simply accumulated vegetation. The coal of our coal beds is very different in its chemical composition from the wood of the trees. It contains a much higher percentage of carbon and a lower percentage of hydrogen and oxygen than ordinary vegetable substances. The conversion of the vegetation of the carboniferous ages into coal was accompanied by a gradual loss of hydrogen and a consequent increase in the percentage of carbon. It is this change that has added to the density of the substance and makes the greater value of coal as fuel. There is little doubt now as to the method by which this woody material of the past has been converted into coal. The same process appears to be going on in a similar manner to-day in the peat beds of various northern countries. The fallen vegetation, trees, trunks, branches, and leaves, accumulate in masses, and, when the conditions of moisture and temperature are right, begin to undergo a fermentation.

Ordinarily this action of bacteria, as already noticed, produces an almost complete though slow oxidation of the carbon, and results in the total decay of the vegetable matter. But if the vegetable mass be covered by water and mud under proper conditions of moisture and temperature, a different kind of fermentation arises which does not produce such complete decay. The covering of water prevents the access of oxygen to the fermenting mass, an oxidation of the carbon is largely prevented, and the vegetable matter slowly changes its character. Under the influence of this slow fermentation, aided, probably by pressure, the mass becomes more and more solid and condensed, its woody character becomes less and less distinct, and there is a gradual loss of the hydrogen and the oxygen. Doubtless there is a loss of carbon also, for there is an evolution of marsh gas which contains carbon. But, in this slow fermentation taking place under the water in peat bogs and marshes the carbon loss is relatively small; the woody material does not become completely oxidized, as it does in free operations of decay. The loss of hydrogen and oxygen from the mass is greater than that of carbon, and the percentage of carbon therefore increases. This is not the ordinary kind of fermentation that goes on in vegetable accumulations. It requires special conditions and possibly special kinds of fermenting organisms. Peat is not formed in all climates. In warm regions, or where the woody matter is freely exposed to the air, the fermentation of vegetable matter is more complete, and it is entirely destroyed by oxidation. It is only in colder regions and when covered with water that the destruction of the organic matter stops short of decay. But such incomplete fermentation is still going on in many parts of the world, and by its means vegetable accumulations are being converted into peat.

This formation of peat appears to be a first step in the formation of denser coal. By a continuation of the same processes the mass becomes still more dense and solid. As we pass from the top to the bottom of such an accumulation of peat, we find it becoming denser and denser, and at the bottom it is commonly of a hard consistence, brownish in colour, and with only slight traces of the original woody structure. Such material is called lignite. It contains a higher percentage of carbon than peat, but a lower percentage than coal, and is plainly a step in coal formation. But the process goes on, the hydrogen and oxygen loss continuing until there is finally produced true coal.

If this is the correct understanding of the formation of coal, we see that we have plainly a process in which bacterial life has had a large and important share. We are, of course, densely ignorant of the exact processes going on. We know nothing positively as to the kind of microorganisms which produce this slow, peculiar fermentation. As yet, the fermentation going on in the formation of the peat has not been studied by the bacteriologists, and we do not know from direct experiment that it is a matter of bacterial action. It has been commonly regarded as simply a slow chemical change, but its general similarity to other fermentative processes is so great that we can have little hesitation in attributing it to micro-organisms, and doubtless to some forms of plants allied to bacteria. There is no reason for doubting that bacteria existed in the geological ages with essentially the same powers as they now possess, and to some forms of bacteria which grow in the absence of oxygen can we probably attribute the slow change which has produced coal. Here, then, is another great source of wealth in Nature for which we are dependent upon bacteria. While, of course, water and pressure were very essential factors in the deposition of coal, it was a peculiar kind of fermentation occurring in the vegetation that brought about the chemical changes in it which resulted in its transformation into coal. The vegetation of the carboniferous age was dependent upon the nitrogen fixed by the bacteria, and to these organisms also do we owe the fact that this vegetation was stored for us in the rocks.

CHAPTER V.

PARASITIC BACTERIA AND THEIR RELATION TO DISEASE.

Perhaps the most universally known fact in regard to bacteria is that they are the cause of disease. It is this fact that has made them objects of such wide interest. This is the side of the subject that first attracted attention, has been most studied, and in regard to which there has been the greatest accumulation of evidence. So persistently has the relation of bacteria to disease been discussed and emphasized that the majority of readers are hardly able to disassociate the two. To most people the very word bacteria is almost equivalent to disease, and the thought of swallowing microbes in drinking water or milk is decidedly repugnant and alarming. In the public mind it is only necessary to demonstrate that an article holds bacteria to throw it under condemnation.

We have already seen that bacteria are to be regarded as agents for good, and that from their fundamental relation to plant life they must be looked upon as our friends rather than as our enemies. It is true that there is another side to the story which relates to the parasitic species. These parasitic forms may do us direct or indirect injury. But the species of bacteria which are capable of doing us any injury, the pathogenic bacteria, are really very few compared to the great host of species which are harmless. A small number of species, perhaps a score or two, are pathogenic, while a much larger number, amounting to hundreds and perhaps thousands of species, are perfectly harmless. This latter class do no injury even though swallowed by man in thousands. They are not parasitic, and are unable to grow in the body of man. Their presence is entirely consistent with the most perfect health, and, indeed, there are some reasons for believing that they are sometimes directly beneficial to health. It is entirely unjust to condemn all bacteria because a few chance to produce mischief. Bacteria in general are agents for good rather than ill.

There are, however, some species which cause mankind much trouble by interfering in one way or another with the normal processes of life. These pathogenic bacteria, or disease germs, do not all act alike, but bring about injury to man in a number of different ways. We may recognise two different classes among them, which, however, we shall see are connected by intermediate types. These two classes are, first, the pathogenic bacteria, which are not strictly parasitic but live free in Nature; and, second, those which live as true parasites in the bodies of man or other animals. To understand the real relation of these two classes, we must first notice the method by which bacteria in general produce disease.

METHOD BY WHICH BACTERIA PRODUCE DISEASE.

Since it was first clearly recognised that certain species of bacteria have the power of producing disease, the question as to how they do so has ever been a prominent one Even if they do grow in the body, why should their presence give rise to the symptoms characterizing disease? Various answers to this question have been given in the past It has been suggested that in their growth they consume the food of the body and thus exhaust it, that they produce an oxidation of the body tissues, or that they produce a reduction of these tissues, or that they mechanically interfere with the circulation None of these suggestions have proved of much value Another view was early advanced, and has stood the test of time. This claim is that the bacteria while growing in the body produce poisons, and these poisons then have a direct action on the body We have already noticed that bacteria during their growth in any medium produce a large number of biproducts of decomposition. We noticed also that among these biproducts there are some which have a poisonous nature; so poisonous are they that when inoculated into the body of an animal they may produce poisoning and death. We have only to suppose that the pathogenic bacteria, when growing as parasites in man, produce such poisons, and we have at once an explanation of the method by which they give rise to disease.

This explanation of germ disease is more than simple theory. It has been in many cases clearly demonstrated. It has been found that the bacteria which cause diphtheria, tetanus, typhoid, tuberculosis, and many other diseases, produce, even when growing in common culture media, poisons which are of a very violent nature. These poisons when inoculated into the bodies of animals give rise to much the same symptoms as the bacteria do themselves when growing as parasites in the animals. The chief difference in the results from inoculating an animal with the poison and with the living bacteria is in the rapidity of the action. When the poison is injected the poisoning symptoms are almost immediately seen, but when the living bacteria are inoculated the effect is only seen after several days or longer, not, in short, until the inoculated bacteria have had time enough to grow in the body and produce the poison in quantity. It has not by any means been shown that all pathogenic germs produce their effect in this way, but it has been proved to be the real method in quite a number of cases, and is extremely probable in others. While some bacteria perhaps produce results by a different method, we must recognise the production of poisons as at all events the common direct cause of the symptoms of disease. This explanation will enable us more clearly to understand the relation of different bacteria to disease.

PATHOGENIC GERMS WHICH ARE NOT STRICTLY PARASITIC

Recognising that bacteria may produce poisons, we readily see that it is not always necessary that they should be parasites in order to produce trouble. In their ordinary growth in Nature such bacteria will produce no trouble The poisons will be produced in decaying material but will seldom be taken into the human body. These poisons, produced in the first stages of putrefaction, are oxidized by further stages of decomposition into harmless products. But should it happen that some of these bacteria obtained a chance to grow vigorously for a while in organic products that are subsequently swallowed as man's food, it is plain that evil results might follow. If such food is swallowed by man after the bacteria have produced their poisonous bodies, it will tend to produce an immediate poisoning of his system. The effect may be sudden and severe if considerable quantity of the poisonous material is swallowed, or slight but protracted if small quantities are repeatedly consumed in food. Such instances are not uncommon. Well-known examples are cases of ice-cream poisoning, poisoning from eating cheese or from drinking milk, or in not a few instances from eating fish or meats within which bacteria have had opportunity for growth. In all these cases the poison is swallowed in quantity sufficient to give rise quickly to severe symptoms, sometimes resulting fatally, and at other times passing off as soon as the body succeeds in throwing off the poisons. In other cases still, however, the amount of poison swallowed may be very slight, too slight to produce much effect unless the same be consumed repeatedly. All such trouble may be attributed to fermented or partly decayed food. It is difficult to distinguish such instances from others produced in a slightly different way, as follows:

It may happen that the bacteria which grow in food products continue to grow in the food even after it is swallowed and has passed into the stomach or intestines. This appears particularly true of milk bacteria. Under these conditions the bacteria are not in any proper sense parasitic, since they are simply living in and feeding upon the same food which they consume outside the body, and are not feeding upon the tissues of man. The poisons which they produce will continue to be developed as long as the bacteria continue to grow, whether in a milk pail or a human stomach. If now the poisons are absorbed by the body, they may produce a mild or severe disease which will be more or less lasting, continuing perhaps as long as the same food and the same bacteria are supplied to the individual. The most important disease of this class appears to be the dreaded cholera infantum, so common among infants who feed upon cow's milk in warm weather. It is easy to understand the nature of this disease when we remember the great number of bacteria in milk, especially in hot weather, and when we remember that the delicate organism of the infant will be thrown at once into disorder by slight amounts of poison which would have no appreciable effect upon the stronger adult. We can easily understand, further, how the disease readily yields to treatment if care is taken to sterilize the milk given to the patient.

We do not know to-day the extent of the troubles which are produced by bacteria of this sort. They will, of course, be chiefly connected with our food products, and commonly, though not always, will affect the digestive functions. It is probable that many of the cases of summer diarrhoea are produced by some such cause, and if they could be traced to their source would be found to be produced by bacterial poisons swallowed with food or drink, or by similar poisons produced by bacteria growing in such food after it is swallowed by the individual. In hot weather, when bacteria are so abundant everywhere and growing so rapidly, it is impossible to avoid such dangers completely without exercising over all food a guard which would be decidedly oppressive. It is well to bear in mind, however, that the most common and most dangerous source of such poisons is milk or its products, and for this reason one should hesitate to drink milk in hot weather unless it is either quite fresh or has been boiled to destroy its bacteria.

PATHOGENIC BACTERIA WHICH ARE TRUE PARASITES.

This class of pathogenic bacteria includes those which actually invade the body and feed upon its tissues instead of living simply upon swallowed food. It is difficult, however, to draw any sharp line separating the two classes. The bacteria which cause diphtheria (Fig. 28), for instance, do not really invade the body. They grow in the throat, attached to its walls, and are confined to this external location or to the superficial tissues. This bacillus is, in short, only found in the mouth and throat, and is practically confined to the so-called false membranes. It never enters any of the tissues of the body, although attached to the mucous membrane. It grows vigorously in this membrane, and there secretes or in some way produces extremely violent poisons. These poisons are then absorbed by the body and give rise to the general symptoms of the disease. Much the same is true of the bacillus which causes tetanus or lockjaw (Fig. 29). This bacillus is commonly inoculated into the flesh of the victim by a wound made with some object which has been lying upon the earth where the bacillus lives. The bacillus grows readily after being inoculated, but it is localized at the point of the wound, without invading the tissue to any extent. It produces, however, during its growth several poisons which have been separated and studied. Among them are some of the most violent poisons of which we have any knowledge. While the bacillus grows in the tissues around the wound it secretes these poisons, which are then absorbed by the body generally. Their poisoning effects produce the violent symptoms of the disease. Of much the same nature is Asiatic cholera. This is caused by a bacillus which is able to grow rapidly in the intestines, feeding perhaps in part on the food in the intestines and perhaps in part upon the body secretions. To a slight extent also it appears to be able to invade the tissues of the body, for the bacilli are found in the walls of the intestines. But it is not a proper parasite, and the fatal disease it produces is the result of the absorption of the poisons secreted in the intestines.

It is but a step from this to the true parasites. Typhoid fever, for example, is a disease produced by bacteria which grow in the intestines, but which also invade the tissues more extensively than the cholera germs (Fig. 30). They do not invade the body generally, however, but become somewhat localized in special glands like the liver, the spleen, etc. Even here they do not appear to find a very favourable condition, for they do not grow extensively in these places. They are likely to be found in the spleen in small groups or centres, but not generally distributed through it. Wherever they grow they produce poison, which has been called typhotoxine, and it is this poison chiefly which gives rise to the fever.

Quite a considerable number of the pathogenic germs are, like the typhoid bacillus, more or less confined to special places. Instead of distributing themselves through the body after they find entrance, they are restricted to special organs. The most common example of a parasite of this sort is the tuberculosis bacillus, the cause of consumption, scrofula, white swelling, lupus, etc. (Fig. 31). Although this bacillus is very common and is able to attack almost any organ in the body, it is usually very restricted in growth. It may become localized in a small gland, a single joint, a small spot in the lungs, or in the glands of the mesentery, the other parts of the body remaining free from infection. Not infrequently the whole trouble is thus confined to such a small locality that nothing serious results. But in other instances the bacilli may after a time slowly or rapidly distribute themselves from these centres, attacking more and more of the body until perhaps fatal results follow in the end. This disease is therefore commonly of very slow progress.

Again, we have still other parasites which are not thus confined, but which, as soon as they enter the body, produce a general infection, attacking the blood and perhaps nearly all tissues simultaneously. The most typical example of this sort is anthrax or malignant pustule, a disease fortunately rare in man (Fig. 32). Here the bacilli multiply in the blood, and very soon a general and fatal infection of the whole body arises, resulting from the abundance of the bacilli everywhere. Some of the obscure diseases known as blood poisoning appear to be of the same general nature, these diseases resulting from a very general invasion of the whole body by certain pathogenic bacteria.

In general, then, we see that the so-called germ diseases result from the action upon the body of poisons produced by bacterial growth. Differences in the nature of these poisons produce differences in the character of the disease, and differences in the parasitic powers of the different species of bacteria produce wide differences in the course of the diseases and their relation to external phenomena.

WHAT DISEASES ARE DUE TO BACTERIA?

It is, of course, an extremely important matter to determine to what extent human diseases are caused by bacteria. It is not easy, nor indeed possible, to do this to-day with accuracy. It is no easy matter to prove that any particular disease is caused by bacteria. To do this it is necessary to find some particular bacterium present in all cases of the disease; to find some method of getting it to grow outside the body in culture media; to demonstrate its absence in healthy animals, or healthy human individuals if it be a human disease; and, finally, to reproduce the disease in healthy animals by inoculating them with the bacterium. All of these steps of proof present difficulties, but especially the last one. In the study of animals it is comparatively easy to reproduce a disease by inoculation. But experiments upon man are commonly impossible, and in the case of human diseases it is frequently very difficult or impossible to obtain the final test of the matter. After finding a specific bacterium associated with a disease, it is usually possible to experiment with it further upon animals only. But some human diseases do not attack animals, and in the case of diseases that may be given to animals it is frequently uncertain whether the disease produced in the animal by such inoculation is identical with the human disease in question, owing to the difference of symptoms in the different animals. As a consequence, the proof of the germ nature of different diseases varies all the way from absolute demonstration to mere suspicion. To give a complete and correct list of the diseases caused by bacteria, or to give a list of the bacteria species pathogenic to man, is therefore at present impossible.

The difficulty of giving such a list is rendered greater from the fact that we have in recent years learned that the same species of pathogenic bacterium may produce different results under different conditions. When the subject of germ disease was first studied and the connection between bacteria and disease was first demonstrated, it was thought that each particular species of pathogenic bacteria produced a single definite disease; and conversely, each germ disease was supposed to have its own definite species of bacterium as its cause. Recent study has shown, however, that this is not wholly true. It is true that some diseases do have such a definite relation to definite bacteria. The anthrax germ, for example, will always produce anthrax, no matter where or how it is inoculated into the body. So, also, in quite a number of other cases distinct specific bacteria are associated with distinct diseases. But, on the other hand, there are some pathogenic bacteria which are not so definite in their action, and produce different results in accordance with circumstances, the effect varying both with the organ attacked and with the condition of the individual. For instance, a considerable number of different types of blood poisoning, septicaemia, pyaemia, gangrene, inflammation of wounds, or formation of pus from slight skin wounds—indeed, a host of miscellaneous troubles, ranging all the way from a slight pus formation to a violent and severe blood poisoning—all appear to be caused by bacteria, and it is impossible to make out any definite species associated with the different types of these troubles. There are three common forms of so-called pus cocci, and these are found almost indiscriminately with various types of inflammatory troubles. Moreover, these species of bacteria are found with almost absolute constancy in and around the body, even in health. They are on the clothing, on the skin, in the mouth and alimentary canal. Here they exist, commonly doing no harm. They have, however, the power of doing injury if by chance they get into wounds. But their power of doing injury varies both with the condition of the individual and with variations in the bacteria themselves. If the individual is in a good condition of health these bacteria have little power of injuring him even when they do get into such wounds, while at times of feeble vitality they may do much more injury, and take the occasion of any little cut or bruise to enter under the skin and give rise to inflammation and pus. Some people will develop slight abscesses or slight inflammations whenever the skin is bruised, while with others such bruises or cuts heal at once without trouble. Both are doubtless subject to the same chance of infection, but the one resists, while the other does not. In common parlance, we say that such a tendency to abscesses indicates a bad condition of the blood—a phrase which means nothing. Further, we find that the same species of bacterium may have varying powers of producing disease at different times. Some species are universal inhabitants of the alimentary canal and are ordinarily harmless, while under other conditions of unknown character they invade the tissues and give rise to a serious and perhaps fatal disease. We may thus recognise some bacteria which may be compared to foreign invaders, while others are domestic enemies. The former, like the typhoid bacillus, always produce trouble when they succeed in entering the body and finding a foothold. The latter, like the normal intestinal bacilli, are always present but commonly harmless, only under special conditions becoming troublesome. All this shows that there are other factors in determining the course of a disease, or even the existence of a disease, than the simple presence of a peculiar species of pathogenic bacterium.

From the facts just stated it will be evident that any list of germ diseases will be rather uncertain. Still, the studies of the last twenty years or more have disclosed some definite relations of bacteria and disease, and a list of the diseases more or less definitely associated with distinct species of bacteria is of interest. Such a list, including only well-known diseases, is as follows:

Name of disease. Name of bacterium producing the disease.
Anthrax (Malignant pustule). Bacillus anthracis.
Cholera. Spirillum cholera: asiaticae
Croupous pneumonia. Micrococcus pneumonia crouposa.
Diphtheria. Bacillus diphtheria.
Glanders. Bacillus mallei.
Gonorrhoea. Micrococcus gonorrhaeae
Influenza. Bacillus of influenza.
Leprosy. Bacillus leprae.
Relapsing fever. Spirillum Obermeieri.
Tetanus (lockjaw). Bacillus tetani.
Tuberculosis (including
consumption, scrofula, etc.) Bacillus tuberculosis.
Typhoid fever. Bacillus typhi abdominalis.

Various wound infections, including septicaemia, pyaemia, acute abscesses, ulcers, erysipelas, etc., are produced by a few forms of micrococci, resembling each other in many points but differing slightly. They are found almost indiscriminately in any of these wound infections, and none of them appears to have any definite relation to any special form of disease unless it be the micrococcus of erysipelas. The common pus micrococci are grouped under three species, Staphylococcus pyogenes aureus, Staphylococcus pyogenes, and Streptococcus pyogenes. These three are the most common, but others are occasionally found.

In addition to these, which may be regarded as demonstrated, the following diseases are with more or less certainty regarded as caused by distinct specific bacteria: Bronchitis, endocarditis, measles, whooping-cough, peritonitis, pneumonia, syphilis.

Still another list might be given of diseases whose general nature indicates that they are caused by bacteria, but in connection with which no distinct bacterium has yet been found. As might be expected also, a larger list of animal diseases has been demonstrated to be caused by these organisms. In addition, quite a number of species of bacteria have been found in such material as faeces, putrefying blood, etc., which have been shown by experiment to be capable of producing diseases in animals, but in regard to which we have no evidence that they ever do produce actual disease under any normal conditions. These may contribute, perhaps, to the troubles arising from poisonous foods, but can not be regarded as disease germs proper.

VARIABILITY OF PATHOGENIC POWERS.

As has already been stated, our ideas of the relation of bacteria to disease have undergone quite a change since they were first formulated, and we recognise other factors influencing disease besides the actual presence of the bacterium. These we may briefly consider under two heads, viz., variation in the bacterium, and variation in the susceptibility of the individual. The first will require only a brief consideration.

That the same species of pathogenic bacteria at different times varies in its powers to produce disease has long been known. Various conditions are known to affect thus the virulence of bacteria. The bacillus which is supposed to give rise to pneumonia loses its power to produce the disease after having been cultivated for a short time in ordinary culture media in the laboratory. This is easily understood upon the suggestion that it is a parasitic bacillus and does not thrive except under parasitic conditions. Its pathogenic powers can sometimes be restored by passing it again through some susceptible animal. One of the most violent pathogenic bacteria is that which produces anthrax, but this loses its pathogenic powers if it is cultivated for a considerable period at a high temperature. The micrococcus which causes fowl cholera loses its power if it be cultivated in common culture media, care being taken to allow several days to elapse between the successive inoculations into new culture flasks. Most pathogenic bacteria can in some way be so treated as to suffer a diminution or complete loss of their powers of producing a fatal disease. On the other hand, other conditions will cause an increase in the virulence of a pathogenic germ. The virus which produces hydrophobia is increased in violence if it is inoculated into a rabbit and subsequently taken from the rabbit for further inoculation. The fowl cholera micrococcus, which has been weakened as just mentioned, may be restored to its original violence by inoculating it into a small bird, like a sparrow, and inoculating a second bird from this. A few such inoculations will make it as active as ever. These variations doubtless exist among the species in Nature as well as in artificial cultures. The bacteria which produce the various wound infections and abscesses, etc., appear to vary under normal conditions from a type capable of producing violent and fatal blood poisoning to a type producing only a simple abscess, or even to a type that is entirely innocuous. It is this factor, doubtless, which in a large measure determines the severity of any epidemic of a bacterial contagious disease.

SUSCEPTIBILITY OF THE INDIVIDUAL.

The very great modification of our early views has affected our ideas as to the power which individuals have of resisting the invasion of pathogenic bacteria. It has from the first been understood that some individuals are more susceptible to disease than others, and in attempting to determine the significance of this fact many valuable and interesting discoveries have been made. After the exposure to the disease there follows a period of some length in which there are no discernible effects. This is followed by the onset of the disease and its development to a crisis, and, if this be passed, by a recovery. The general course of a germ disease is divided into three stages: the stage of incubation, the development of the disease, and the recovery. The susceptibility of the body to a disease may be best considered under the three heads of Invasion, Resistance, Recovery.

Means of Invasion.—In order that a germ disease should arise in an individual, it is first necessary that the special bacterium which causes the disease should get into the body. There are several channels through which bacteria can thus find entrance; these are through the mouth, through the nose, through the skin, and occasionally through excretory ducts. Those which come through the mouth come with the food or drink which we swallow; those which enter through the nose must be traced to the air; and those which enter through the skin come in most cases through contact with some infected object, such as direct contact with the body of an infected person or his clothing or some objects he has handled, etc. Occasionally, perhaps, the bacteria may get into the skin from the air, but this is certainly uncommon and confined to a few diseases. There are here two facts of the utmost importance for every one to understand: first, that the chance of disease bacteria being carried to us through the air is very slight and confined to a few diseases, such as smallpox, tuberculosis, scarlet fever; etc., and, secondly, that the uninjured skin and the uninjured mucous membrane also is almost a sure protection against the invasion of the bacteria. If the skin is whole, without bruises or cuts, bacteria can seldom, if ever, find passage through it. These two facts are of the utmost importance, since of all sources of infection we have the least power to guard against infection through the air, and since of all means 'of entrance we can guard the skin with the greatest difficulty. We can easily render food free from pathogenic bacteria by heating it. The material we drink can similarly be rendered harmless, but we can not by any known means avoid breathing air, nor is there any known method of disinfecting the air, and it is impossible for those who have anything to do with sick persons to avoid entirely having contact either with the patient or with infected clothing or utensils.

From the facts here given it will be seen that the individual's susceptibility to disease produced by parasitic bacteria will depend upon his habits of cleanliness, his care in handling infectious material, or care in cleansing the hands after such handling, upon his habit of eating food cooked or raw, and upon the condition of his skin and mucous membranes, since any kind of bruises will increase susceptibility. Slight ailments, such as colds, which inflame the mucous membrane, will decrease its resisting power and render the individual more susceptible to the entrance of any pathogenic germs should they happen to be present. Sores in the mouth or decayed teeth may in the same way be prominent factors in the individual's susceptibility. Thus quite a number of purely physical factors may contribute to an individual's susceptibility.

Resisting Power of the Body.—Even after the bacteria get into the body it is by no means certain that they will give rise to disease, for they have now a battle to fight before they can be sure of holding their own. It is now, indeed, that the actual conflict between the powers of the body and these microscopic invaders begins. After they have found entrance into the body the bacteria have arrayed against them strong resisting forces of the human organism, endeavouring to destroy and expel them. Many of them are rapidly killed, and sometimes they are all destroyed without being able to gain a foothold. In such cases, of course, no trouble results. In other cases the body fails to overcome the powers of the invaders and they eventually multiply rapidly. In this struggle the success of the invaders is not necessarily a matter of numbers. They are simply struggling to gain a position in the body, where they can feed and grow. A few individuals may be entirely sufficient to seize such a foothold, and then these by multiplying may soon become indefinitely numerous. To protect itself, therefore, the human body must destroy every individual bacterium, or at least render them all incapable of growth. Their marvellous reproductive powers give the bacteria an advantage in the battle. On the other hand, it takes time even for these rapidly multiplying beings to become sufficiently numerous to do injury. There is thus an interval after their penetration into the body when these invaders are weak in numbers. During this interval—the period of incubation—the body may organize a resistance sufficient to expel them.

We do not as yet thoroughly understand the forces which the human organism is able to array against these invading foes. Some of its methods of defence are, however, already intelligible to us, and we know enough, at all events, to give us an idea of the intensity of the conflict that is going on, and of the vigorous and powerful forces which the human organism is able to bring against its invading enemies.

In the first place, we notice that a majority of bacteria are utterly unable to grow in the human body even if they do find entrance. There are known to bacteriologists to-day many hundreds, even thousands of species, but the vast majority of these find in the human tissues conditions so hostile to their life that they are utterly unable to grow therein. Human flesh or human blood will furnish excellent food for them if the individual be dead, but living human flesh and blood in some way exerts a repressing influence upon them which is fatal to the growth of a vast majority of species. Some few species, however, are not thus destroyed by the hostile agencies of the tissues of the animal, but are capable of growing and multiplying in the living body. These alone are what constitute the pathogenic bacteria, since, of course, these are the only bacteria which can produce disease by growing in the tissues of an animal. The fact that the vast majority of bacteria can not grow in the living organism shows clearly enough that there are some conditions existing in the living tissue hostile to bacterial life. There can be little doubt, moreover, that it is these same hostile conditions, which enable the body to resist the attack of the pathogenic species in cases where resistance is successfully made.

What are the forces arrayed against these invaders? The essential nature of the battle appears to be a production of poisons and counter poisons. It appears to be an undoubted fact that the first step in repelling these bacteria is to flood them with certain poisons which check their growth. In the blood and lymph of man and other animals there are present certain products which have a direct deleterious influence upon the growth of micro-organisms. The existence of these poisons is undoubted, many an experiment having directly attested to their presence in the blood of animals. Of their nature we know very little, but of their repressing influence upon bacterial growth we are sure. They have been named alexines, and they are produced in the living tissue, although as to the method of their production we are in ignorance. By the aid of these poisons the body is able to prevent the growth of the vast majority of bacteria which get into its tissues. Ordinary micro-organisms are killed at once, for these alexines act as antiseptics, and common bacteria can no more grow in the living body than they could in a solution containing other poisons Thus the body has a perfect protection against the majority of bacteria. The great host of species which are found in water, milk, air, in our mouths or clinging to our skin, and which are almost omnipresent in Nature, are capable of growing well enough in ordinary lifeless organic foods, but just as soon as they succeed in finding entrance into living human tissue their growth is checked at once by these antiseptic agents which are poured upon them. Such bacteria are therefore not pathogenic germs, and not sources of trouble to human health.

There are, on the other hand, a few species of bacteria which may be able to retain their lodgment in the body m spite of this attempt of the individual to get rid of them. These, of course, constitute the pathogenic species, or so called "disease germs". Only such species as can overcome this first resistance can be disease germs, for they alone can retain their foothold in the body.

But how do these species overcome the poisons, which kill the other harmless bacteria? They, as well as the harmless forms, find these alexines injurious to their growth, but in some way they are able to counteract the poisons. In this general discussion of poisons we are dealing with a subject which is somewhat obscure, but apparently the pathogenic bacteria are able to overcome the alexines of the body by producing in their turn certain other products which neutralize the alexines, thus annulling their action. These pathogenic bacteria, when they get into the body, give rise at once to a group of bodies which have been named lysines. These lysines are as mysterious to us as the alexines, but they neutralize the effect of the alexines and thus overcome the resistance the body offers to bacterial growth The invaders can now multiply rapidly enough to get a lasting foothold in the body and then soon produce the abnormal symptoms which we call disease Pathogenic bacteria thus differ from the non-pathogenic bacteria primarily in this power of secreting products which can neutralize the ordinary effects of the alexines, and so overcome the body's normal resistance to their parasitic life.

Even if the bacteria do thus overcome the alexines the battle is not yet over, for the individual has another method of defence which is now brought into activity to check the growth of the invading organisms. This second method of resistance is by means of a series of active cells found in the blood, known as white blood-corpuscles (Fig. 33 a, b). They are minute bits of protoplasm present in the blood and lymph in large quantities. They are active cells, capable of locomotion and able to crawl out of the blood-vessels Not infrequently they are found to take into their bodies small objects with which they come in contact. One of their duties is thus to engulf minute irritating bodies which may be in the tissues, and to carry them away for excretion. They thus act as scavengers These corpuscles certainly have some agency in warding off the attacks of pathogenic bacteria Very commonly they collect in great numbers in the region of the body where invading bacteria are found. Such invading bacteria exist upon them a strong attraction, and the corpuscles leave the blood-vessels and sometimes form a solid phalanx completely surrounding the invading germs. Their collection at these points may make itself seen externally by the phenomenon we call inflammation.

There is no question that the corpuscles engage in conflict with the bacteria when they thus surround them. There has been not a little dispute, however, as to the method by which they carry on the conflict. It has been held by some that the corpuscles actually take the bacteria into their bodies, swallow them, as it were, and subsequently digest them (Fig. 33 c, d, e). This idea gave rise to the theory of phagocytosis, and the corpuscles were consequently named phagocytes. The study of several years has, however, made it probable that this is not the ordinary method by which the corpuscles destroy the bacteria. According to our present knowledge the method is a chemical one. These cells, when they thus collect in quantities around the invaders, appear to secrete from their own bodies certain injurious products which act upon the bacteria much as do the alexines already mentioned. These new bodies have a decidedly injurious effect upon the multiplying bacteria; they rapidly check their growth, and, acting in union with the alexines, may perhaps entirely destroy them.

After the bacteria are thus killed, the white blood-corpuscles may load themselves with their dead bodies and carry them away (Fig. 33 d, e). Sometimes they pass back into the blood stream and carry the bacteria to various parts of the body for elimination. Not infrequently the white corpuscles die in the contest, and then may accumulate in the form of pus and make their way through the skin to be discharged directly. The battle between these phagocytes and the bacteria goes on vigorously. If in the end the phagocytes prove too strong for the invaders, the bacteria are gradually all destroyed, and the attack is repelled. Under these circumstances the individual commonly knows nothing—of the matter. This conflict has taken place entirely without any consciousness on his part, and he may not even know that he has been exposed to the attack of the bacteria. In other cases the bacteria prove too strong for the phagocytes. They multiply too rapidly, and sometimes they produce secretions which actually drive the phagocytes away. Commonly, as already noticed, the corpuscles are attracted to the point of invasion, but in some cases, when a particularly deadly and vigorous species of bacteria invades the body, the secretions produced by them are so powerful as actually to drive the corpuscles away. Under these circumstances the invading hosts have a chance to multiply unimpeded, to distribute themselves over the body, and the disease rapidly follows as the result of their poisoning action on the body tissues.

It is plain, then, that the human body is not helpless in the presence of the bacteria of disease, but that it is supplied with powerful resistant forces. It must not be supposed, however, that the outline of the action of these forces just given is anything like a complete account of the matter; nor must it be inferred that the resistance is in all respects exactly as outlined. The subject has only recently been an object of investigation, and we are as yet in the dark in regard to many of the facts. The future may require us to modify to some extent even the brief outline which has been given. But while we recognise this uncertainty in the details, we may be assured of the general facts. The living body has some very efficacious resistant forces which prevent most bacteria from growing within its tissues, and which in large measure may be relied upon to drive out the true pathogenic bacteria. These resistant forces are in part associated with the productions of body poisons, and are in part associated with the active powers of special cells which have been called phagocytes. The origin of the poisons and the exact method of action of the phagocytes we may well leave to the future to explain.

These resisting powers of the body will vary with conditions. It is evident that they are natural powers, and they will doubtless vary with the general condition of vigour of the individual. Robust health, a body whose powers are strong, well nourished, and vigorous, will plainly furnish the conditions for the greatest resistance to bacterial diseases. One whose bodily activities are weakened by poor nutrition can offer less resistance. The question whether one shall suffer from a germ disease is not simply the question whether he shall be exposed, or even the question whether the bacteria shall find entrance into his body. It is equally dependent upon whether he has the bodily vigour to produce alexines in proper quantity, or to summon the phagocytes in sufficient abundance and vigour to ward off the attack. We may do much to prevent disease by sanitation, which aids in protecting the individual from attack; but we must not forget that the other half of the battle is of equal importance, and hence we must do all we can to strengthen the resisting forces of the organism.

RECOVERY FROM GERM DISEASES.

These resisting forces are not always sufficient to drive off the invaders. The organisms may retain their hold in the body for a time and eventually break down the resistance. After this they may multiply unimpeded and take entire possession of the body. As they become more numerous their poisonous products increase and begin to produce direct poisoning effects on the body. The incubation period is over and the disease comes on. The disease now runs its course. It becomes commonly more and more severe until a crisis is reached. Then, unless the poisoning is so severe that death occurs, the effects pass away and recovery takes place.

But why should not a germ disease be always fatal? If the bacteria thus take possession of the body and can grow there, why do they not always continue to multiply until they produce sufficient poison to destroy the life of the individual? Such fatal results do, of course, occur, but in by far the larger proportion of cases recovery finally takes place.

Plainly, the body must have another set of resisting forces which is concerned in the final recovery. Although weakened by the poisoning and suffering from the disease, it does not yield the battle, but somewhat slowly organizes a new attack upon the invaders. For a time the multiplying bacteria have an unimpeded course and grow rapidly; but finally their further increase is checked, their vigour impaired, and after this they diminish in numbers and are finally expelled from the body entirely. Of the nature of this new resistance but little is yet known. We notice, in the first place, that commonly after such a recovery the individual has decidedly increased resistance to the disease. This increased resistance may be very lasting, and may be so considerable as to give almost complete immunity from the disease for many years, or for life. One attack of scarlet fever gives the individual great immunity for the future. On the other hand, the resistance thus derived may be very temporary, as in the case of diphtheria. But a certain amount of resistance appears to be always acquired. This power of resisting the activities of the parasites seems to be increased during the progress of the disease, and, if it becomes sufficient, it finally drives off the bacteria before they have produced death. After this, recovery takes place. To what this newly acquired resisting power is due is by no means clear to bacteriologists, although certain factors are already known. It appears beyond question that in the case of certain diseases the cells of the body after a time produce substances which serve as antidotes to the poisons produced by the bacteria during their growth in the body-antitoxines. In the case of diphtheria, for instance, the germs growing in the throat produce poisons which are absorbed by the body and give rise to the symptoms of the disease; but after a time the body cells react, and themselves produce a counter toxic body which neutralizes the poisonous effect of the diphtheria poison. This substance has been isolated from the blood of animals that have recovered from an attack of diphtheria, and has been called diphtheria antitoxine. But even with this knowledge the recovery is not fully explained. This antitoxine neutralizes the effects of the diphtheria toxine, and then the body develops strength to drive off the bacteria which have obtained lodgment in the throat. How they accomplish this latter achievement we do not know as yet. The antitoxme developed simply neutralizes the effects of the toxine. Some other force must be at work to get rid of the bacteria, a force which can only exert itself after the poisoning effect of the poison is neutralized. In these cases, then, the recovery is due, first, to the development in the body of the natural antidotes to the toxic poisons, and, second, to some other unknown force which drives off the parasites.

These facts are certainly surprising. If one had been asked to suggest the least likely theory to explain recovery from disease, he could hardly have found one more unlikely than that the body cells developed during the disease an antidote to the poison which the disease bacteria were producing. Nevertheless, it is beyond question that such antidotes are formed during the course of the germ diseases. It has not yet been shown in all diseases, and it would be entirely too much to claim that this is the method of recovery in all cases. We may say, however, in regard to bacterial diseases in general, that after the bacteria enter the body at some weak point they have first a battle to fight with the resisting powers of the body, which appear to be partly biological and partly chemical. These resisting powers are in many cases entirely sufficient to prevent the bacteria from obtaining a foothold. If the invading host overcome the resisting powers, then they begin to multiply rapidly, and take possession of the body or some part of it. They continue to grow until either the individual dies or something occurs to check their growth. After the individual develops the renewed powers of checking their growth, recovery takes place, and the individual is then, because of these renewed powers of resistance, immune from a second attack of the disease for a variable length of time.

This, in the merest outline, represents the relation of bacterial parasites to the human body But while this is a fair general expression of the matter, it must be recognised that different diseases differ much in their relations, and no general outline will apply to all They differ in their method of attack and in the point of attack. Not only do they produce different kinds of poisons giving rise to different symptoms of poisoning; not only do they produce different results in different animals; not only do the different pathogenic species differ much in their power to develop serious disease, but the different species are very particular as to what species of animal they attack. Some of them can live as parasites in man alone; some can live as parasites upon man and the mouse and a few other animals; some can live in various animals but not in man; some appear to be able to live in the field mouse, but not in the common mouse; some live in the horse; some in birds, but not in warm-blooded mammals; while others, again, can live almost equally well in the tissues of a long list of animals. Those which can live as parasites upon man are, of course, especially related to human disease, and are of particular interest to the physician, while those which live in animals are in a similar way of interest to veterinarians.

Thus we see that parasitic bacteria show the widest variations. They differ in point of attack, in method of attack, and in the part of the body which they seize upon as a nucleus for growth. They differ in violence and in the character of the poisons they produce, as well as in their power of overcoming the resisting powers of the body. They differ at different times in their powers of producing disease. In short, they show such a large number of different methods of action that no general statements can be made which will apply universally, and no one method of guarding against them or in driving them off can be hoped to apply to any extended list of diseases.

DISEASES CAUSED BY OTHER ORGANISMS THAN BACTERIA.

Although the purpose of this work is to deal primarily with the bacterial world, it would hardly be fitting to leave the subject without some reference to diseases caused by organisms which do not belong to the group of bacteria. While most of the so-called germ diseases are caused by the bacteria which we have been studying in the previous chapters, there are some whose inciting cause is to be found among organisms belonging to other groups. Some of these are plants of a higher organization than bacteria, but others are undoubtedly microscopic animals. Their life habits are somewhat different from those of bacteria, and hence the course of the diseases is commonly different. Of the diseases thus produced by microscopic animals or by higher plants, one or two are of importance enough to deserve special mention here.

Malaria.—The most important of these diseases is malaria in its various forms, and known under various names—chills and fever, autumnal fever, etc. This disease, so common almost everywhere, has been studied by physicians and scientists for a long time, and many have been the causes assigned to it. At one time it was thought to be the result of the growth of a bacterium, and a distinct bacillus was described as producing it. It has finally been shown, however, to be caused by a microscopic organism belonging to the group of unicellular animals, and somewhat closely related to the well-known amoeba. This organism is shown in Fig. 34. The whole history of the malarial organism is not yet known. The following statements comprise the most important facts known in regard to it, and its relation to the disease in man.

Undoubtedly the malarial germ has some home outside the human body, but it is not yet very definitely known what this external home is; nor do we know from what source the human parasite is derived. It appears probable that water serves in some cases as its means of transference to man, and air in other cases. From some external source it gains access to man and finds its way into the blood. Here it attacks the red blood-corpuscles, each malarial organism making its way into a single one (Fig. 340). Here it now grows, increasing in size at the expense of the substance of the corpuscle. As it becomes larger it becomes granular, and soon shows a tendency to separate into a number of irregular masses. Finally it breaks up into many minute bodies called spores. These bodies break out of the corpuscle and for a time live a free life in the blood. After a time they make their way into other red blood-corpuscles, develop into new malarial amoeboid parasites, and repeat the growth and sporulation. This process can apparently be repeated many times without check.

These organisms are thus to be regarded as parasites of the red corpuscles. It is, of course, easy to believe that an extensive parasitism and destruction of the corpuscles would be disastrous to the health of the individual, and the severity of the disease will depend upon the extent of the parasitism. Corresponding to this life history of the organism, the disease malaria is commonly characterized by a decided intermittency, periods of chill and fever alternating with periods of intermission in which these symptoms are abated. The paroxysms of the disease, characterized by the chill, occur at the time that the spores are escaping from the blood-corpuscles and floating in the blood. After they have again found their way into a blood-corpuscle the fever diminishes, and during their growth in the corpuscle until the next sporulation the individual has a rest from the more severe symptoms.

There appears to be more than one variety of the malarial organism, the different types differing in the length of time it takes for their growth and sporulation. There is one variety, the most common one, which requires two days for its growth, thus giving rise to the paroxysm of the disease about once in forty- eight hours; another variety appears to require three days for its growth; while still another variety appears to be decidedly irregular in its period of growth and sporulation. These facts readily explain some of the variations in the disease. Certain other irregularities appear to be due to a different cause. More than one brood of parasites may be in the blood of the individual at the same time, one producing sporulation at one time and another at a different time. Such a simultaneous growth of two independent broods may plainly produce almost any kind of modification in the regularity of the disease.

The malarial organism appears to be very sensitive to quinine, a very small quantity being sufficient to kill it. Upon this point depends the value of quinine as a medicine. If the drug be present in the blood at the time when the spores are set free from the blood-corpuscle, they are rapidly killed by it before they have a chance to enter another corpuscle. During their growth in the corpuscle they are far less sensitive to quinine than when they exist in the free condition as spores, and at this time the drug has little effect.

The malarial organism is an animal, and can not be cultivated in the laboratory by any artificial method yet devised. Its whole history is therefore not known. It doubtless has some home outside the blood of animals, and very likely it may pass through other stages of a metamorphosis in the bodies of other animals. Most parasitic animals have two or more hosts upon which they live, alternating from one to the other, and that such is the case with the malarial parasite is at least probable. But as yet bacteriologists have been unable to discover anything very definite in regard to the matter. Until we can learn something in regard to its life outside the blood of man we can do little in the way of devising methods to avoid it.

Malaria differs from most germ diseases in the fact that the organisms which produce it are not eliminated from the body in any way. In most germ diseases the germs are discharged from the patient by secretions or excretions of some kind, and from these excretions may readily find their way into other individuals. The malarial organism is not discharged from the body in any way, and hence is not contagious. If the parasite does pass part of its history in some other animal than man, there must be some means by which it passes from man to its other host. It has been suggested that some of the insects which feed upon human blood may serve as the second host and become inoculated when feeding upon such blood. This has been demonstrated with startling success in regard to the mosquito (Anopheles), some investigators going so far as to say that this is the only way in which the disease can be communicated.

Several other microscopic animals occur as parasites upon man, and some of them are so definitely associated with certain diseases as to lead to the belief that they are the cause of these diseases. The only one of very common occurrence is a species known as Amaeba coli, which is found in cases of dysentery. In a certain type of dysentery this organism is so universally found that there is little doubt that it is in some very intimate way associated with the cause of the disease. Definite proof of the matter is, however, as yet wanting.

On the side of plants, we find that several plants of a higher organization than bacteria may become parasitic upon the body of man and produce various types of disease. These plants belong mostly to the same group as the moulds, and they are especially apt to attack the skin. They grow in the skin, particularly under the hair, and may send their threadlike branches into some of the subdermal tissues. This produces irritation and inflammation of the skin, resulting in trouble, and making sores difficult to heal. So long as the plant continues to grow, the sores, of course, can not be healed, and when the organisms get into the skin under the hair it is frequently difficult to destroy them. Among the diseases thus caused are ringworm, thrush, alopecia, etc.

CHAPTER VI.

METHODS OF COMBATING PARASITIC BACTERIA.

The chief advantage of knowing the cause of disease is that it gives us a vantage ground from which we may hope to find means of avoiding its evils. The study of medicine in the past history of the world has been almost purely empirical, with a very little of scientific basis. Great hopes are now entertained that these new facts will place this matter upon a more strictly scientific foundation. Certainly in the past twenty-five years, since bacteriology has been studied, more has been done to solve problems connected with disease than ever before. This new knowledge has been particularly directed toward means of avoiding disease. Bacteriology has thus far borne fruit largely in the line of preventive medicine, although to a certain extent also along the line of curative medicine. This chapter will be devoted to considering how the study of bacteriology has contributed directly and indirectly to our power of combating disease.

PREVENTIVE MEDICINE.

In the study of medicine in the past centuries the only aim has been to discover methods of curing disease; at the present time a large and increasing amount of study is devoted to the methods of preventing disease. Preventive medicine is a development of the last few years, and is based almost wholly upon our knowledge of bacteria. This subject is yearly becoming of more importance. Forewarned is forearmed, and it has been found that to know the cause of a disease is a long step toward avoiding it. As some of our contagious and epidemic diseases have been studied in the light of bacteriological knowledge, it has been found possible to determine not only their cause, but also how infection is brought about, and consequently how contagion may be avoided. Some of the results which have grown up so slowly as to be hardly appreciated are really great triumphs. For instance, the study of bacteriology first led us to suspect, and then demonstrated, that tuberculosis is a contagious disease, and from the time that this was thus proved there has been a slow, but, it is hoped, a sure decline in this disease. Bacteriological study has shown that the source of cholera infection in cases of raging epidemics is, in large part at least, our drinking water; and since this has been known, although cholera has twice invaded Europe, and has been widely distributed, it has not obtained any strong foothold or given rise to any serious epidemic except in a few cases where its ravages can be traced to recognised carelessness. It is very significant to compare the history of the cholera epidemics of the past few years with those of earlier dates. In the epidemics of earlier years the cholera swept ruthlessly through communities without check. In the last few years, although it has repeatedly knocked at the doors of many European cities, it has been commonly confined to isolated cases, except in a few instances where these facts concerning the relation to drinking water were ignored.

The study of preventive medicine is yet in its infancy, but it has already accomplished much. It has developed modern systems of sanitation, has guided us in the building of hospitals, given rules for the management of the sick-room which largely prevent contagion from patient to nurse; it has told us what diseases are contagious, and in what way; it has told us what sources of contagion should be suspected and guarded against, and has thus done very much to prevent the spread of disease. Its value is seen in the fact that there has been a constant decrease in the death rate since modern ideas of sanitation began to have any influence, and in the fact that our general epidemics are less severe than in former years, as well as in the fact that more people escape the diseases which were in former times almost universal.

The study of preventive medicine takes into view several factors, all connected with the method and means of contagion. They are the following:

The Source of Infectious Material.—t has been learned that for most diseases the infectious material comes from individuals suffering with the disease, and that except in a few cases, like malaria, we must always look to individuals suffering from disease for all sources of contagion. It is found that pathogenic bacteria are in all these cases eliminated from the patient in some way, either from the alimentary canal or from skin secretions or otherwise, and that any nurse with common sense can have no difficulty in determining in what way the infectious material is eliminated from her patients. When this fact is known and taken into consideration it is a comparatively easy matter to devise valuable precautions against distribution of such material. It is thus of no small importance to remember that the simple presence of bacteria in food or drink is of no significance unless these bacteria have come from some source of disease infection.

The Method of Distribution. — The bacteria must next get from the original source of the disease to the new susceptible individual. Bacteria have no independent powers of distribution unless they be immersed in liquids, and therefore their passage from individual to individual must be a passive one. They are readily transferred, however, by a number of different means, and the study of these means is aiding much in checking contagion Study along this line has shown that the means by which bacteria are carried are several. First we may notice food as a distributor. Food may become contaminated by infectious material in many ways; for example, by contact with sewage, or with polluted water, or even with eating utensils which have been used by patients. Water is also likely to be contaminated with infectious material, and is a fertile source for distributing typhoid and cholera. Milk may become contaminated in a variety of ways, and be a source of distributing the bacteria which produce typhoid fever, tuberculosis, diphtheria, scarlet fever, and a few other less common diseases. Again, infected clothing, bedding, or eating utensils may be taken from a patient and be used by another individual without proper cleansing. Direct contact, or contact with infected animals, furnishes another method. Insects sometimes carry the bacteria from person to person, and in some diseases (tuberculosis, and perhaps scarlet fever and smallpox) we must look to the air as a distributor of the infectious material. Knowledge of these facts is helping to account for multitudes of mysterious cases of infection, especially when we combine them with the known sources of contagious matter.

Means of Invasion.—Bacteriology has shown us that different species of parasitic bacteria have different means of entering the body, and that each must enter the proper place in order to get a foothold. After we learn that typhoid infectious material must enter the mouth in order to produce the disease; that tuberculosis may find entrance through the nose in breathing, while types of blood poisoning enter only through wounds or broken skin, we learn at once fundamental facts as to the proper methods of meeting these dangers. We learn that with some diseases care exercised to prevent the swallowing of infectious material is sufficient to prevent contagion, while with others this is entirely insufficient. When all these facts are understood it is almost always perfectly possible to avoid contagion; and as these facts become more and more widely known direct contagion is sure to become less frequent.

Above all, it is telling us what becomes of the pathogenic bacteria after being eliminated from the body of the patient; how they may exist for a long time still active; how they may lurk in filth or water dormant but alive, or how they may even multiply there. Preventive medicine is telling us how to destroy those thus lying in wait for a chance of infection, by discovering disinfectants and telling us especially where and when to use them. It has already taught us how to crush out certain forms of epidemics by the proper means of destroying bacteria, and is lessening the dangers from contagious diseases. In short, the study of bacteriology has brought us into a condition where we are no longer helpless in the presence of a raging epidemic. We no longer sit in helpless dismay, as did our ancestors, when an epidemic enters a community, but, knowing their causes and sources, set about at once to remove them. As a result, severe epidemics are becoming comparatively short-lived.

BACTERIA IN SURGERY.

In no line of preventive medicine has bacteriology been of so much value and so striking in its results as in surgery. Ever since surgery has been practised surgeons have had two difficulties to contend with. The first has been the shock resulting from the operation. This is dependent upon the extent of the operation, and must always be a part of a surgical operation. The second has been secondary effects following the operation. After the operation, even though it was successful, there were almost sure to arise secondary complications known as surgical fever, inflammation, blood poisoning, gangrene, etc., which frequently resulted fatally. These secondary complications were commonly much more serious than the shock of the operation, and it used to be the common occurrence for the patient to recover entirely from the shock, but yield to the fevers which followed. They appeared to be entirely unavoidable, and were indeed regarded as necessary parts of the healing of the wound. Too frequently it appeared that the greater the care taken with the patient the more likely he was to suffer from some of these troubles. The soldier who was treated on the battlefield and nursed in an improvised field hospital would frequently recover, while the soldier who had the fortune to be taken into the regular hospital, where greater care was possible, succumbed to hospital gangrene. All these facts were clearly recognised, but the surgeon, through ignorance of their cause, was helpless in the presence of these inflammatory troubles, and felt it always necessary to take them into consideration.

The demonstration that putrefaction and decay were caused by bacteria, and the early proof that the silkworm disease was produced by a micro-organism, led to the suggestion that the inflammatory diseases accompanying wounds were similarly caused. There are many striking similarities between these troubles and putrefaction, and the suggestion was an obvious one. At first, however, and for quite a number of years, it was impossible to demonstrate the theory by finding the distinct species of micro- organisms which produced the troubles. We have already seen that there are several different species of bacteria which are associated with this general class of diseases, but that no specific one has any particular relation to a definite type of inflammation. This fact made discoveries in this connection a slow matter from the microscopical standpoint. But long before this demonstration was finally reached the theory had received practical application in the form of what has developed into antiseptic or aseptic surgery.

Antiseptic surgery is based simply upon the attempt to prevent the entrance of bacteria into the surgical wound. It is assumed that if these organisms are kept from the wound the healing will take place without the secondary fevers and inflammations which occur if they do get a chance to grow in the wound. The theory met with decided opposition at first, but accumulating facts demonstrated its value, and to-day its methods have been adopted everywhere in the civilized world. As the evidence has been accumulating, surgeons have learned many important facts, foremost among which is a knowledge of the common sources from which the infection of wounds occurs. At first it was thought that the air was the great source of infection, but the air bacteria have been found to be usually harmless. It has appeared that the more common sources are the surgeon's instruments, or his hands, or the clothing or sponges which are allowed to come in contact with the wounds. It has also appeared that the bacteria which produce this class of troubles are common species, existing everywhere and universally present around the body, clinging to the clothing or skin, and always on hand to enter the wound if occasion offers. They are always present, but commonly harmless. They are not foreign invaders like the more violent pathogenic species, such as those of Asiatic cholera, but may be compared to domestic enemies at hand. It is these ever-present bacteria which the surgeon must guard against. The methods by which he does this need not detain us here. They consist essentially in bacteriological cleanliness. The operation is performed with sterilized instruments under most exacting conditions of cleanliness.

The result has been a complete revolution in surgery. As the methods have become better understood and more thoroughly adopted, the instances of secondary troubles following surgical wounds have become less and less frequent until they have practically disappeared in all simple cases. To-day the surgeon recognises that when inflammatory troubles of this sort follow simple surgical wounds it is a testimony to his carelessness. The skilful surgeon has learned that with the precautions which he is able to take to-day he has to fear only the direct effect of the shock of the wound and its subsequent direct influence; but secondary surgical fevers, blood poisoning, and surgical gangrene need not be taken into consideration at all. Indeed, the modern surgeon hardly knows what surgical gangrene is, and bacteriologists have had practically no chance to study it. Secondary infections have largely disappeared, and the surgeon is concerned simply with the effect of the wound itself, and the power of the body to withstand the shock and subsequently heal the wound.

With these secondary troubles no longer to disturb him, the surgeon has become more and more bold. Operations formerly not dreamed of are now performed without hesitation. In former years an operation which opened the abdominal cavity was not thought possible, or at least it was so nearly certain to result fatally that it was resorted to only on the last extremity; while to-day such operations are hardly regarded as serious. Even brain surgery is becoming more and more common. Possibly our surgeons are passing too far to the other extreme, and, feeling their power of performing so many operations without inconvenience or danger, they are using the knife in cases where it would be better to leave Nature to herself for her own healing. But, be this as it may, it is impossible to estimate the amount of suffering prevented and the number of lives saved by the mastery of the secondary inflammatory troubles which used to follow surgical wounds.

Preventive medicine, then, has for its object the prevention rather than the cure of disease. By showing the causes of disease and telling us where and how they are contracted, it is telling us how they may to a large extent be avoided. Unlike practical medicine, this subject is one which has a direct relation to the general public. While it may be best that the knowledge of curative methods be confined largely to the medical profession, it is eminently desirable that a knowledge of all the facts bearing upon preventive medicine should be distributed as widely as possible. One person can not satisfactorily apply his knowledge of preventive medicine, if his neighbour is ignorant of or careless of the facts. We can not hope to achieve the possibilities lying along this line until there is a very wide distribution of knowledge. Every epidemic that sweeps through our communities is a testimony to the crying need of education in regard to such simple facts as the source of infectious material, the methods of its distribution, and the means of rendering it harmless.

PREVENTION IN INOCULATION.

It has long been recognised that in most cases recovery from one attack of a contagious disease renders an individual more or less immune against a second attack. It is unusual for an individual to have the same contagious disease twice. This belief is certainly based upon fact, although the immunity thus acquired is subject to wide variations. There are some diseases in which there is little reason for thinking that any immunity is acquired, as in the case of tuberculosis, while there are others in which the immunity is very great and very lasting, as in the case of scarlet fever. Moreover, the immunity differs with individuals. While some persons appear to acquire a lasting immunity by recovery from a single attack, others will yield to a second attack very readily. But in spite of this the fact of such acquired immunity is beyond question. Apparently all infectious diseases from which a real recovery takes place are followed by a certain amount of protection from a second attack; but with some diseases the immunity is very fleeting, while with others it is more lasting. Diseases which produce a general infection of the whole system are, as a rule, more likely to give rise to a lasting immunity than those which affect only small parts. Tuberculosis, which, as already noticed, is commonly quite localized in the body, has little power of conveying immunity, while a disease like scarlet fever, which affects the whole system, conveys a more lasting protection.

Such immunity has long been known, and in the earlier years was sometimes voluntarily acquired; even to-day we find some individuals making use of the principle. It appears that a mild attack of such diseases produces immunity equally well with a severe attack, and acting upon this fact mothers have not infrequently intentionally exposed their children to certain diseases at seasons when they are mild, in order to have the disease "over with" and their children protected in the future. Even the more severe diseases have at times been thus voluntarily acquired. In China it has sometimes been the custom thus to acquire smallpox. Such methods are decidedly heroic, and of course to be heartily condemned. But the principle that a mild type of the disease conveys protection has been made use of in a more logical and defensible way.

The first instance of this principle was in vaccination against smallpox, now practised for more than a century. Cowpox is doubtless closely related to smallpox, and an attack of the former conveys a certain amount of protection against the latter. It was easy, therefore, to inoculate man with some of the infectious material from cowpox, and thus give him some protection against the more serious smallpox. This was a purely empirical discovery, and vaccination was practised long before the principle underlying it was understood, and long before the germ nature of disease was recognised. The principle was revived again, however, by Pasteur, and this time with a logical thought as to its value. While working upon anthrax among animals, he learned that here, as in other diseases, recovery, when it occurred, conveyed immunity. This led him to ask if it were not possible to devise a method of giving to animals a mild form of the disease and thus protect them from the more severe type. The problem of giving a mild type of this extraordinarily severe disease was not an easy one. It could not be done, of course, by inoculating the animals with a small number of the bacteria, for their power of multiplication would soon make them indefinitely numerous. It was necessary in some way to diminish their violence. Pasteur succeeded in doing this by causing them to grow in culture fluids for a time at a high temperature. This treatment diminished their violence so much that they could be inoculated into cattle, where they produced only the mildest type of indisposition, from which the animals speedily recovered. But even this mild type of the disease was triumphantly demonstrated to protect the animals from the most severe form of anthrax. The discovery was naturally hailed as a most remarkable one, and one which promised great things in the future. If it was thus possible, by direct laboratory methods, to find a means of inoculating against a serious disease like anthrax, why could not the same principle be applied to human diseases? The enthusiasts began at once to look forward to a time when all diseases should be thus conquered.

But the principle has not borne the fruit at first expected. There is little doubt that it might be applied to quite a number of human diseases if a serious attempt should be made. But several objections arise against its wide application. In the first place, the inoculation thus necessary is really a serious matter. Even vaccination, as is well known, sometimes, through faulty methods, results fatally, and it is a very serious thing to experiment upon human beings with anything so powerful for ill as pathogenic bacteria. The seriousness of the disease smallpox, its extraordinary contagiousness, and the comparatively mild results of vaccination, have made us willing to undergo vaccination at times of epidemics to avoid the somewhat great probability of taking the disease. But mankind is unwilling to undergo such an operation, even though mild, for the purpose of avoiding other less severe diseases, or diseases which are less likely to be taken. We are unwilling to be inoculated against mild diseases, or against the more severe ones which are uncommon. For instance, a method has been devised for rendering animals immune against lockjaw, which would probably apply equally well to man. But mankind in general will never adopt it, since the danger from lockjaw is so small. Inoculation must then be reserved for diseases which are so severe and so common, or which occur in periodical epidemics of so great severity, as to make people in general willing to submit to inoculation as a protection. A further objection arises from the fact that the immunity acquired is not necessarily lasting. The cattle inoculated against anthrax retain their protective powers for only a few months. How long similar immunity might be retained in other cases we can not say, but plainly this fact would effectually prevent this method of protecting mankind from being used except in special cases. It is out of the question to think of constant and repeated inoculations against various diseases.

As a result, the principle of inoculation as an aid in preventive medicine has not proved of very much value. The only other human disease in which it has been attempted seriously is Asiatic cholera. This disease in times of epidemics is so severe and the chance of infection is so great as to justify such inoculation. Several bacteriologists have in the last few years been trying to discover a harmless method of inoculating against this disease. Apparently they have succeeded, for experiments in India, the home of the cholera, have been as successful as could be anticipated. Bacteriological science has now in its possession a means of inoculation against cholera which is perhaps as efficacious as vaccination is against smallpox. Whether it will ever be used to any extent is doubtful, since, as already pointed out, we are in a position to avoid cholera epidemics by other means. If we can protect our communities by guarding the water supply, it is not likely that the method of inoculation will ever be widely used.

Another instance of the application of preventive inoculation has been made, but one based upon a different principle. Hydrophobia is certainly one of the most horrible of diseases, although comparatively rare. Its rarity would effectually prevent mankind from submitting to a general inoculation against it, but its severity would make one who had been exposed to it by the bite of a rabid animal ready to submit to almost any treatment that promised to ward off the disease. In the attempt to discover a means of inoculating against this disease it was necessary, therefore, to find a method that could be applied after the time of exposure—i.e., after the individual had been bitten by the rabid animal. Fortunately, the disease has a long period of incubation, and one that has proved long enough for the purpose. A method of inoculation against this disease has been devised by Pasteur, which can be applied after the individual has been bitten by the rabid animal. Apparently, however, this preventive inoculation is dependent upon a different principle from vaccination or inoculation against anthrax. It does not appear to give rise to a mild form of the disease, thus protecting the individual, but rather to an acquired tolerance of the chemical poisons produced by the disease. It is a well-known physiological fact that the body can become accustomed to tolerate poisons if inured to them by successively larger and larger doses. It is by this power, apparently, that the inoculation against hydrophobia produces its effect. Material containing the hydrophobia poison (taken from the spinal cord of a rabbit dead with the disease) is injected into the individual after he has been bitten by a rabid animal. The poisonous material in the first injection is very weak, but is followed later by a more powerful inoculation. The result is that after a short time the individual has acquired the power of resisting the hydrophobia poisons. Before the incubation period of the original infectious matter from the bite of the rabid animal has passed, the inoculated individual has so thoroughly acquired a tolerance of the poison that he successfully resists the attack of the infection. This method of inoculation thus neutralizes the effects of the disease by anticipating them.

The method of treatment of hydrophobia met with extraordinarily violent opposition. For several years it was regarded as a mistake. But the constantly accumulating statistics from the Pasteur Institute have been so overwhelmingly on one side as to quiet opposition and bring about a general conviction that the method is a success.

The method of preventive inoculation has not been extensively applied to human diseases in addition to those mentioned. In a few cases a similar method has been used to guard against diphtheria. Among animals, experiment has shown that such methods can quite easily be obtained, and doubtless the same would be true of mankind if it was thought practical or feasible to apply them. But, for reasons mentioned, this feature of preventive medicine will always remain rather unimportant, and will be confined to a few of the more violent diseases.

It may be well to raise the question as to why a single attack with recovery conveys immunity. This question is really a part of the one already discussed as to the method by which the body cures disease. We have seen that this is in part due to the development of chemical substances which either neutralize the poisons or act as germicide upon the bacteria, or both, and perhaps due in part to an active destruction of bacteria by cellular activity (phagocytosis). There is little reason to doubt that it is the same set of activities which renders the animal immune. The forces which drive off the invading bacteria in one case are still present to prevent a second attack of the same species of bacterium. The length of time during which these forces are active and sufficient to cope with any new invaders determines the length of time during which the immunity lasts. Until, therefore, we can answer with more exactness just how cure is brought about in case of disease, we shall be unable to explain the method of immunity.

LIMITS OF PREVENTIVE MEDICINE.

With all the advance in preventive medicine we can not hope to avoid disease entirely. We are discovering that the sources of disease are on all sides of us, and so omnipresent that to avoid them completely is impossible. If we were to apply to our lives all the safeguards which bacteriology has taught us should be applied in order to avoid the different diseases, we would surround ourselves with conditions which would make life intolerable. It would be oppressive enough for us to eat no food except when it is hot, to drink no water except when boiled, and to drink no milk except after sterilization; but these would not satisfy the necessary conditions for avoiding disease. To meet all dangers, we should handle nothing which has not been sterilized, or should follow the handling by immediately sterilizing the hands; we should wear only disinfected clothes, we should never put our fingers in our mouths or touch our food with them; we should cease to ride in public conveyances, and, indeed, should cease to breathe common air. Absolute prevention of the chance of infection is impossible. The most that preventive medicine can hope for is to point out the most common and prolific sources of infection, and thus enable civilized man to avoid some of his most common troubles. It becomes a question, therefore, where we will best draw the line in the employment of safeguards. Shall we drink none except sterilized milk, and no water unless boiled? or shall we put these occasional sources of danger in the same category with bicycle and railroad accidents, dangers which can be avoided by not using the bicycle or riding on the rail, but in regard to which the remedy is too oppressive for application?

Indeed, when viewed in a broad philosophical light it may not be the best course for mankind to shun all dangers. Strength in the organism comes from the use rather than the disuse of our powers. It is certain that the general health and vigour of mankind is to be developed by meeting rather than by shunning dangers. Resistance to disease means bodily vigour, and this is to be developed in mankind by the application of the principle of natural selection. In accordance with this principle, disease will gradually remove the individuals of weak resisting powers, leaving those of greater vigour. Parasitic bacteria are thus a means of preventing the continued life of the weaker members of the community, and so tend to strengthen mankind. By preventive medicine many a weak individual who would otherwise succumb earlier in the struggle is enabled to live a few years longer. Whatever be our humanitarian feeling for the individual, we can not fail to admit that this survival of the weak is of no benefit to the race so far as the development of physical nature is concerned. Indeed, if we were to take into consideration simply the physical nature of man we should be obliged to recommend a system such as the ancient Spartans developed, of exposing to death all weakly individuals, that only the strong might live to become the fathers of future generations. In this light, of course, parasitic diseases would be an assistance rather than a detriment to the human race. Of course such principles will never again be dominant among men, and our conscience tells us to do all we can to help the weak. We shall doubtless do all possible to develop preventive medicine in order to guard the weak against parasitic organisms. But it is at all events well for us to remember that we can never hope to develop the strength of the human race by shunning evil, but rather by combating it, and the power of the human race to resist the invasions of these organisms will never be developed by the line of action which guards us from attack. Here, as in other directions, the principles of modern humanity have, together with their undoubted favourable influence upon mankind, certain tendencies toward weakness. While we shall still do our utmost to develop preventive medicine in a proper way, it may be well for us to remember these facts when we come to the practical question of determining where to draw the limits of the application of methods for preventing infectious diseases.

CURATIVE MEDICINE.

Bacteriology has hitherto contributed less to curative than to preventive medicine. Nevertheless, its contributions to curative medicine have not been unimportant, and there is promise of much more in the future. It is, of course, unsafe to make predictions for the future, but the accomplishments of the last few years give much hope as to further results.

DRUGS.

It was at first thought that a knowledge of the specific bacteria which cause a disease would give a ready means of finding specific drugs for the cure of such disease. If a definite species of bacterium causes a disease and we can cultivate the organism in the laboratory, it is easy to find some drugs which will be fatal to its growth, and these same drugs, it would seem, should be valuable as medicines in these diseases. This hope has, however, proved largely illusive. It is very easy to find some drug which proves fatal to the specific germs while growing in the culture media of the laboratory, but commonly these are of little or no use when applied as medicines. In the first place, such substances are usually very deadly poisons. Corrosive sublimate is a substance which destroys all pathogenic germs with great rapidity, but it is a deadly poison, and can not be used as a drug in sufficient quantity to destroy the parasitic bacteria in the body without at the same time producing poisonous effects on the body itself. It is evident that for any drug to be of value in thus destroying bacteria it must have some specially strong action upon the bacteria. Its germicide action on the bacteria should be so strong that a dose which would be fatal or very injurious to them would be too small to have a deleterious influence on the body of the individual. It has not proved an easy task to discover drugs which will have any value as germicides when used in quantities so small as to produce no injurious effect on the body.

A second difficulty is in getting the drug to produce its effect at the right point. A few diseases, as we have noticed, are produced by bacteria which distribute themselves almost indiscriminately over the body; but the majority are somewhat definitely localized in special points. Tuberculosis may attack a single gland or a single lobe of the lung. Typhoid germ is localized in the intestines, liver, spleen, etc. Even if it were possible to find some drug which would have a very specific effect upon the tuberculosis bacillus, it is plain that it would be a very questionable method of procedure to introduce this into the whole system simply that it might have an effect upon a very small isolated gland. Sometimes such a bacterial affection may be localized in places where it can be specially treated, as in the case of an attack on a dermal gland, and in these cases some of the germicides have proved to be of much value. Indeed, the use of various disinfectants connected with abscesses and superficial infections has proved of much value. To this extent, in disinfecting wounds and as a local application, the development of our knowledge of disinfectants has given no little aid to curative medicine.

Very little success, however, has resulted in the attempt to find specific drugs for specific diseases, and it is at least doubtful whether many such will ever be found. The nearest approach to it is quinine as a specific poison for malarial troubles. Malarious diseases are not, however, produced by bacteria but by a microscopic organism of a very different nature, thought to be an animal rather than a plant. Besides this there has been little or no success in discovering specifics in the form of drugs which can be given as medicines or inoculated with the hope of destroying special kinds of pathogenic bacteria without injury to the body. While it is unwise to make predictions as to future discoveries, there seems at present little hope for a development of curative medicine along these lines.

VIS MEDICATRIX NATURAE.

The study of bacterial diseases as they progress in the body has emphasized above all things the fact that diseases are eventually cured by a natural rather than by an artificial process. If a pathogenic bacterium succeeds in passing the outer safeguards and entering the body, and if it then succeeds in overcoming the forces of resistance which we have already noticed, it will begin to multiply and produce mischief. This multiplication now goes on for a time unchecked, and there is little reason to expect that we can ever do much toward checking it by means of drugs. But after a little, conditions arise which are hostile to the further growth of the parasite. These hostile conditions are produced perhaps in part by the secretions from the bacteria, for bacteria are unable to flourish in a medium containing much of their own secretions. The secretions which they produce are poisons to them as well as to the individual in which they grow, and after these have become quite abundant the further growth of the bacterium is checked and finally stopped. Partly, also, must we conclude that these hostile conditions are produced by active vital powers in the body of the individual attacked. The individual, as we have seen, in some cases develops a quantity of some substance which neutralizes the bacterial poisons and thus prevents their having their maximum effect. Thus relieved from the direct effects of the poisons, the resisting powers are recuperated and once more begin to produce a direct destruction of the bacteria. Possibly the bacteria, being now weakened by the presence of their own products of growth, more readily yield to the resisting forces of the cell life of the body. Possibly the resisting forces are decidedly increased by the reactive effect of the bacteria and their poisons. But, at all events, in cases where recovery from parasitic diseases occurs, the revived powers of resistance finally overcome the bacteria, destroy them or drive them off, and the body recovers.

All this is, of course, a natural process. The recovery from a disease produced by the invasion of parasitic bacteria depends upon whether the body can resist the bacterial poisons long enough for the recuperation of its resisting powers. If these poisons are very violent and produced rapidly, death will probably occur before the resisting powers are strong enough to drive off the bacteria. In the case of some diseases the poisons are so violent that this practically always occurs, recovery being very exceptional. The poison produced by the tetanus bacillus is of this nature, and recovery from lockjaw is of the rarest occurrence. But in many other diseases the body is able to withstand the poison, and later to recover its resisting powers sufficiently to drive off the invaders. In all cases, however, the process is a natural one and dependent upon the vital activity of the body. It is based at the foundation, doubtless, upon the powers of the body cells, either the phagocytes or other active cells. The body has, in short, its own forces for repelling invasions, and upon these forces must we depend for the power to produce recovery.

It is evident that all these facts give us very little encouragement that we shall ever be able to cure diseases directly by means of drugs to destroy bacteria, but, on the contrary, that we must ever depend upon the resisting powers of the body. They teach us, moreover, along what line we must look for the future development of curative medicine. It is evident that scientific medicine must turn its attention toward the strengthening and stimulating of the resisting and curative forces of the body. It must be the physician's aim to enable the body to resist the poisons as well as possible and to stimulate it to re-enforce its resistant forces. Drugs have a place in medicine, of course, but this place is chiefly to stimulate the body to react against its invading hosts. They are, as a rule, not specific against definite diseases. We can not hope for much in the way of discovering special medicines adapted to special diseases. We must simply look upon them as means which the physician has in hand for stimulating the natural forces of the body, and these may doubtless vary with different individual natures. Recognising this, we can see also the logic of the small dose as compared to the large dose. A small dose of a drug may serve as a stimulant for the lagging forces, while a larger dose would directly repress them or produce injurious secondary effects. As soon as we recognise that the aim of medicine is not to destroy the disease but rather to stimulate the resisting forces of the body, the whole logic of therapeutics assumes a new aspect.

Physicians have understood this, and, especially in recent years, have guided their practice by it. If a moderate dose of quinine will check malaria in a few days, it does not follow that twice the dose will do it in half the time or with twice the certainty. The larger doses of the past, intended to drive out the disease, have been everywhere replaced by smaller doses designed to stimulate the lagging body powers. The modern physician makes no attempt to cure typhoid fever, having long since learned his inability to do this, at least if the fever once gets a foothold; but he turns his attention to every conceivable means of increasing the body's strength to resist the typhoid poison, confident that if he can thus enable the patient to resist the poisoning effects of the typhotoxine his patient will in the end react against the disease and drive off the invading bacteria. The physician's duty is to watch and guard, but he must depend upon the vital powers of his patient to carry on alone the actual battle with the bacterial invaders.

ANTITOXINES.

In very recent times, however, our bacteriologists have been pointing out to the world certain entirely new means of assisting the body to fight its battles with bacterial diseases. As already noticed, one of the primal forces in the recovery, from some diseases, at least, is the development in the body of a substance which acts as an antidote to the bacterial poison. So long as this antitoxine is not present the poisons produced by the disease will have their full effect to weaken the body and prevent the revival of its resisting powers to drive off the bacteria. Plainly, if it is possible to obtain this antitoxine in quantity and then inoculate it into the body when the toxic poisons are present, we have a means for decidedly assisting the body in its efforts to drive off the parasites. Such an antidote to the bacterial poison would not, indeed, produce a cure, but it would perhaps have the effect of annulling the action of the poisons, and would thus give the body a much greater chance to master the bacteria. It is upon this principle that is based the use of antitoxines in diphtheria and tetanus

It will be clear that to obtain the antitoxine we must depend upon some natural method for its production. We do not know enough of the chemical nature of the antitoxines to manufacture them artificially. Of course we can not deny the possibility of their artificial production, and certain very recent experiments indicate that perhaps they may be made by the agency of electricity. At present, however, we must use natural methods, and the one commonly adopted is simple. Some animal is selected whose blood is harmless to man and that is subject to the disease to be treated. For diphtheria a horse is chosen. This animal is inoculated with small quantities of the diphtheria poison without the diphtheria bacillus. This poison is easily obtained by causing the diphtheria bacillus to grow in common media in the laboratory for a while, and the toxines develop in quantity; then, by proper filtration, the bacteria themselves can be removed, leaving a pure solution of the toxic poison. Small quantities of this poison are inoculated into the horse at successive intervals. The effect on the horse is the same as if the animal had the disease. Its cells react and produce a considerable quantity of the antitoxine which remains in solution in the blood of the animal. This is not theory, but demonstrated fact. The blood of a horse so treated is found to have the effect of neutralizing the diphtheria poison, although the blood of the horse before such treatment has no such effect. Thus there is developed in the horse's blood a quantity of the antitoxine, and now it may be used by physicians where needed. If some of this horse's blood, properly treated, be inoculated into the body of a person who is suffering from diphtheria, its effect, provided the theory of antitoxines is true, will be to counteract in part, at least, the poisons which are being produced in the patient by the diphtheria bacillus. This does not cure the disease nor in itself drive off the bacilli, but it does protect the body from the poisons to such an extent as to enable it more readily to assert its own resisting powers.

This method of using antitoxines as a help in curing disease is very recent, and we can not even guess what may come of it. It has apparently been successfully applied in diphtheria. It has also been used in tetanus with slight success. The same principle has been used in obtaining an antidote for the poison of snake bites, since it has appeared that in this kind of poisoning the body will develop an antidote to the poison if it gets a chance. Horses have been treated in the same way as with the diphtheria poison, and in the same way they develop a substance which neutralizes the snake poison. Other diseases are being studied to-day with the hope of similar results. How much further the principle will go we can not say, nor can we be very confident that the same principle will apply very widely. The parasitic diseases are so different in nature that we can hardly expect that a method which is satisfactory in meeting one of the diseases will be very likely to be adapted to another. Vaccination has proved of value in smallpox, but is not of use in other human diseases. Inoculation with weakened germs has proved of value in anthrax and fowl cholera, but will not apply to all diseases. Each of these parasites must be fought by special methods, and we must not expect that a method that is of value in one case must necessarily be of use elsewhere. Above all, we must remember that the antitoxines do not cure in themselves; they only guard the body from the weakening effects of the poisons until it can cure itself, and, unless the body has resisting powers, the antitoxine will fail to produce the desired results.

One further point in the action of the antitoxines must be noticed. As we have seen, a recovery from an attack of most germ diseases renders the individual for a time immune against a second attack. This applies less, however, to a recovery after the artificial inoculation with antitoxine than when the individual recovers without such aid. If the individual recovers quite independently of the artificial antitoxine, he does so in part because he has developed the antitoxines for counteracting the poison by his own powers. His cellular activities have, in other words, been for a moment at least turned in the direction of production of antitoxines. It is to be expected, therefore, that after the recovery they will still have this power, and so long as they possess it the individual will have protection from a second attack. When, however, the recovery results from the artificial inoculation of antitoxine the body cells have not actively produced antitoxine. The neutralization of the poisons has been a passive one, and after recovery the body cells are no more engaged in producing antitoxine than before. The antitoxine which was inoculated is soon eliminated by secretion, and the body is left with practically the same liability to attack as before. Its immunity is decidedly fleeting, since it was dependent not upon any activity on the part of the body, but upon an artificial inoculation of a material which is rapidly eliminated by secretion.

CONCLUSION.

It is hoped that the outline which has been given of the bacterial life of Nature may serve to give some adequate idea of these organisms and correct the erroneous impressions in regard to them which are widely prevalent. It will be seen that, as our friends, bacteria play a vastly more important part in Nature than they do as our enemies. These plants are minute and extraordinarily simple, but, nevertheless, there exists a large number of different species. The number of described forms already runs far into the hundreds, and we do not yet appear to be approaching the end of them. They are everywhere in Nature, and their numbers are vast beyond conception. Their powers of multiplication are inconceivable, and their ability to produce profound chemical changes is therefore unlimited. This vast host of living beings thus constitutes a force or series of forces of tremendous significance. Most of the vast multitude we must regard as our friends. Upon them the farmer is dependent for the fertility of his soil and the possibility of continued life in his crops. Upon them the dairyman is dependent for his flavours. Upon them important fermentative industries are dependent, and their universal powers come into action upon a commercial scale in many a place where we have little thought of them in past years. We must look upon them as agents ever at work, by means of which the surface of Nature is enabled to remain fresh and green. Their power is fundamental, and their activities are necessary for the continuance of life. A small number of the vast host, a score or two of species, unfortunately for us, find their most favourable living place in the human body, and thus become human parasites. By their growth they develop poisons and produce disease. This small class of parasites are then decidedly our enemies. But, taken all together, we must regard the bacteria as friends and allies. Without them we should not have our epidemics, but without them we should not exist. Without them it might be that some individuals would live a little longer, if indeed we could live at all. It is true that bacteria, by producing disease, once in a while cause the premature death of an individual; once in a while, indeed, they may sweep off a hundred or a thousand individuals; but it is equally true that without them plant and animal life would be impossible on the face of the earth.

www.ingramcontent.com/pod-product-compliance
Lightning Source LLC
Chambersburg PA
CBHW031422210526
45464CB00005B/2005